SpringerBriefs in Biochemistry and Molecular Biology

For further volumes:
http://www.springer.com/series/10196

Dmitri B. Papkovsky · Alexander V. Zhdanov
Andreas Fercher · Ruslan I. Dmitriev
James Hynes

Phosphorescent
Oxygen-Sensitive
Probes

 Springer

Dmitri B. Papkovsky
Department of Biochemistry
University College Cork
Cork
Ireland

Ruslan I. Dmitriev
Department of Biochemistry
University College Cork
Cork
Ireland

Alexander V. Zhdanov
Department of Biochemistry
University College Cork
Cork
Ireland

James Hynes
Luxcel Biosciences Ltd.
BioInnovation Centre, UCC
Cork
Ireland

Andreas Fercher
Department of Biochemistry
University College Cork
Cork
Ireland

ISSN 2211-9353
ISBN 978-3-0348-0524-7
DOI 10.1007/978-3-0348-0525-4
Springer Basel Heidelberg New York Dordrecht London

ISSN 2211-9361 (electronic)
ISBN 978-3-0348-0525-4 (eBook)

Library of Congress Control Number: 2012945485

Printed on acid-free paper

Springer is part of Springer Science+Business Media (www.springer.com)

Preface

Measurement of dissolved oxygen concentration in biological samples by luminescence quenching method has been introduced by the pioneering work of German scientists Dietrich Lubbers and Norbert Opitz who developed in the mid-1970s first solid-state fluorescence-based O_2 sensors called 'optodes', and by the group of David Wilson in the US who introduced the phosphorescence-based probes and O_2 imaging technique in the mid-1980s. High application potential of this technique has been recognised back then, however its use was rather limited, mainly by research groups who had access to or developed themselves dedicated materials and instrumentation for O_2 sensing and possessed special skills.

In the last decade we have witnessed a major change in the uptake of optical O_2 sensing techniques, with many new materials, measurement formats, detection and imaging platforms, analytical methodologies and accessory tools developed and tested. O_2 sensing systems have been adapted for use with standard laboratory equipment such as time-resolved and lifetime-based luminescent readers, live cell imaging systems, liquid handling equipment (microplates, biochips). They have been applied to various measurement tasks and mechanistic studies with complex biological models demonstrating high utility for biomedical research and new insights into cell and tissue function. O_2 sensor technology has now become much more accessible and affordable for ordinary users working in various disciplines of life and biomedical sciences.

On the other hand, the existence of many probes, measurement formats, and detection platforms make it difficult for the user to select optimal combination to address their particular biological problem or measurement task. Also, distinctive features of these probes compared to other fluorescent probes, and general conditions of their use for O_2 monitoring are not always assessed comprehensively by their end-users. This often leads to experimental artifacts, failures, or incorrect interpretation of data. Critical literature describing their practical uses is also in short supply.

This book is aimed to address these aspects and provide a general overview of existing and emerging O_2 sensing probes, detection platforms, and applications in their various modifications, based on authors' long-standing experience in this area. In the first chapter, the most popular phosphorescent probes based on Pt-porphyrin dyes are described and cross-compared. Subsequently, core biological applications of these probes with different in vitro models (in vivo applications such as imaging of tissue oxygen are outside the scope) are described. For these applications, which are divided into two main groups and chapters—plate reader analysis and O_2 imaging—key technical details are provided on how to set them up, conduct the measurements, extract the analytical and physiological information, interpret the results, and perform troubleshooting. Altogether, this gives potential users a fair representation of merits and limitations, analytical capabilities of the different probes, O_2 sensing and imaging platform(s), and a comprehensive practical guide for their rational selection. The book is expected to facilitate a broader use of the probes and development of new applications.

Dmitri B. Papkovsky

Contents

Chapter 1
O$_2$-Sensitive Probes Based on Phosphorescent Metalloporphyrins

Ruslan I. Dmitriev and Dmitri B. Papkovsky

Abstract Measurement of molecular O$_2$ in biological samples represents an important group of analytical methods actively employed in diverse areas of biology (microbes, plants, animals), medicine and toxicology. In this chapter, the significance, classification of main methods and principles of quenched-phosphorescence measurements of O$_2$ with the help of metalloporphyrin based probes are described. Various measurement platforms are discussed with particular attention to the experimental models.

Keywords Oxygen-sensitive probes · Pt-porphyrins · Phosphorescence quenching · Time-resolved fluorescence · Intracellular probes · Oxygen sensing and imaging · In vitro assays · Cellular oxygen

1.1 Introduction

Measurement of molecular oxygen (O$_2$) in biological samples containing respiring cells and tissues is of high practical and biomedical importance. O$_2$ is a small, gaseous, non-polar analyte which has moderate solubility in aqueous media (~ 200 μM at air saturation, 37 °C). It is supplied to cells and tissues by passive diffusion and, in higher multicellular organisms, by convectional transport via vasculature, red blood cells and haemoglobin [1, 2]. In mammalian cells O$_2$ is the key metabolite and the source of energy involved in the production of ATP through the electron transport chain and oxidative phosphorylation [3]. It is a substrate of numerous enzymatic reactions vital for cellular function, involved in cell signalling and genetic adaptation to hypoxia [4, 5]. Therefore, detailed understanding of biological roles of O$_2$ is of fundamental importance for cell biology, medicine, drug discovery and other disciplines [1, 6].

D. B. Papkovsky et al., *Phosphorescent Oxygen-Sensitive Probes*,
SpringerBriefs in Biochemistry and Molecular Biology,
DOI: 10.1007/978-3-0348-0525-4_1, © The Author(s) 2012

The main analytical tasks in O$_2$ measurement are: (1) assessment of bulk oxygenation of samples containing cells, tissues, organs and whole organisms; (2) measurement of O$_2$ consumption rates (OCR); (3) analysis of O$_2$ distribution, localised gradients and O$_2$ maps in heterogeneous samples; (4) analysis of sub-cellular O$_2$ gradients, and (5) monitoring of dynamics of parameters 1–4 in response to changes in cellular function, for example, in normal/resting and diseased/stimulated cells and tissues.

Analytical task 1 probably has the highest importance: under normal physiological conditions, O$_2$ levels in different tissues are maintained within the defined limits, which are tissue-specific [7, 8]. Significant alterations in oxygenation from the norm are observed in diseased tissues and under pathological conditions, e.g. in solid tumours, under ischaemia/stroke, anaemia, neurodegeneration, hypertension, metabolic disorders. Short-term and sustained hypoxia can lead to cell death or protection and adaptive responses via rearrangement of cell metabolism. The latter includes Warburg effect, hypoxia-induced expression of regulatory genes and proteins such as HIF-1α, PGC-1α [2, 5, 9–11] and their downstream products. On the other hand, significant spatial and temporal fluctuations in O$_2$ occur in exercised skeletal and cardiac muscles, excited regions of the brain, kidney during their normal function [2, 12–14].

OCR reflects respiratory activity of a sample and, together with other biomarkers such as ATP content, mitochondrial membrane potential, metabolite concentrations and ion fluxes, provides important information on the metabolic activity and bioenergetic status. Significant alterations in cellular OCR reflect perturbed metabolism, mitochondrial dysfunction, disease state or drug toxicity [3]. Analytical tasks 3–5 are best addressed by means of O$_2$ imaging techniques, which allow mapping of O$_2$ concentration within biological samples in 2D, 3D and 4D (time lapse experiments), and with sub-micrometer spatial resolution.

Due to the high importance of O$_2$ measurement, the diversity of analytical tasks and biological objects to be analysed, different measurement methodologies have been developed for O$_2$ sensing. The particular sample, measurement location, concentration range, sampling frequency and resolution of O$_2$ to be measured determine the choice of experimental technique, measurement format, detection modality, the particular probe, instrumentation and other tools.

Among these the following main groups can be defined:

1. Electrochemical methods utilising electrodes, such as the Clark electrode system.
2. Physical methods utilising paramagnetic properties of O$_2$, such as EPR spectrometry.
3. Optical methods.

The Clark-type electrode [15] has a relatively simple set-up and low cost. In this system, sample O$_2$ diffuses through a Teflon membrane to a Pt electrode polarised at about +0.7 V against the Ag/AgCl electrode, where it gets reduced generating current proportional to O$_2$ concentration. Its main applications are point measurement of dissolved O$_2$ and absolute OCR of macroscopic biological samples containing suspension cells or isolated mitochondria in a sealed, stirred and temperature controlled cuvette, as well as measurement of local O_2 levels in cells and tissues with microelectrodes [16, 17]. Their principal drawbacks are O_2 consumption by the electrode itself, the need for stirring and regular maintenance, baseline drift, poisoning and fouling of the electrode, poor compatibility with adherent cells and effector treatments, limited throughput. In the past years, much progress has been made in addressing these limitations and adapting the technology for use with adherent cell lines in multiparametric biochips [16] or customised systems for cerebellar granule neurons [17, 18].

Electron paramagnetic resonance (EPR) is a physical method for detecting molecules with unpaired electrons. Since O_2 is a paramagnetic molecule, EPR can be used for its quantification, directly or indirectly using dedicated probes. Extracellular EPR probes such as India Ink [19] were developed for clinical use and assessment of cell populations. Nitroxyl and esterified trityl radicals (e.g. triarylmethyl) represent promising probes for EPR imaging of intracellular O_2 [20, 21]. In vivo EPR imaging with micron resolution ($30 \times 30 \times 100$ µm) has been demonstrated [22] which complements the other O_2 sensing techniques well. EPR spectra and intensity signals can be used for quantification.

Optical methods rely on endogenous or exogenous probes which alter their properties in response to fluctuations in O_2 concentration. The absorption-based methods (e.g. myoglobin in muscle tissue [23]) have been complemented by the luminescence-based techniques which include the measurement of fluorescence of redox indicators (e.g. NADH and FAD) [24], delayed fluorescence of endogenous protoporphyrin IX [25, 26], photoacoustic tissue imaging [27], GFP-based biosensor constructs [28–30] and quenched-luminescence O_2 sensing [31–33].

Luminescence quenching represents one of the most powerful and versatile techniques which allows direct, minimally or non-invasive, real-time monitoring and imaging of O_2 in biological samples with good selectivity and tunable sensitivity [32, 34]. This technology provides reliable and accurate detection of O_2 in different formats including single point macroscopic sensors and microsensors, in vitro bioassays based on O_2 detection, screening platforms (cell, enzyme and animal based) operating on commercial fluorescent readers, sophisticated live cell and in vivo imaging systems, multi-parametric systems in which O_2 detection is coupled with the other probes and biomarkers. A number of such systems and applications have already gained wide use and are produced commercially [35, 36].

The key components of the optical O_2 sensing technique are dedicated luminescent materials that enable O_2 to be probed in complex biological objects, particularly those containing respiring cells. The main O_2 pools that require quantification and monitoring are: (i) dissolved extracellular O_2 in growth medium or vasculature; (ii) pericellular O_2 in the interstitial space, at cell surface; (iii) intracellular O_2 in the cytosol, mitochondria or other compartments; (iv) in vivo measurement of O_2 distribution in live tissues, organs and whole organisms.

Pt(II)- and Pd(II)-porphyrins and some related structures possessing strong phosphorescence at room temperature, moderate quenchability by O_2 and high chemical stability are among the common indicator dyes used in O_2-sensing materials [32, 37]. Fluorescent complexes of Ru(II), Os(II), Ir(III)-porphyrins, lanthanide chelates are also being actively explored in O$_2$ sensor chemistries [38–40], however, they are outside the scope of this book. In this chapter, we will focus on the general principles of quenched-luminescence detection of O$_2$, main types of sensor materials on the basis of phosphorescent porphyrin dyes, different measurement formats of O$_2$ sensing technique and detection modalities and their core application in conjunction with various biological models.

1.2 Principles of Quenched-Phosphorescence Detection of O$_2$

Molecular O$_2$ is an efficient quencher of long-lived excited triplet states which acts via collisional interaction with luminophore molecules causing their radiationless deactivation and return to the ground state. In the presence of O$_2$ phosphorescence intensity (I) and lifetime (τ) are both reduced, the relationship between measured luminescent parameter and O$_2$ concentration is described by the Stern–Volmer equation [41]:

$$I_0/I = \tau_0/\tau = 1 + K_{s-v*}[O_2] = 1 + k_{q*}\tau_{0*}[O_2], \qquad (1.1)$$

where I$_0$ and τ_0 are unquenched intensity and lifetime at zero O$_2$, respectively, K$_{s-v}$ is the Stern–Volmer quenching constant, and k$_q$—the bimolecular quenching rate constant, which depends on the immediate environment of the reporter dye, temperature and sterical factors. Each luminescent material has a characteristic relationship between [O$_2$] and τ (or I). Luminescence lifetime represents the average time which the luminophore stays in the excited state before emitting a photon. This is an intrinsic feature of the material independent on the dye concentration and measurement set-up. For this reason, lifetime is a preferred measurement parameter for O$_2$ quantification by luminescence quenching. Microsecond range lifetimes of the phosphorescent dyes are relatively easy to measure, unlike the nanosecond lifetimes of conventional fluorophores which require short excitation pulses and high speed detectors [42]. By conducting phosphorescence lifetime or intensity measurements with the sensor, O$_2$ concentration within test sample can be quantified as follows:

$$[O_2] = (\tau_0 - \tau)/(\tau_* K_{s-v}) \qquad (1.2)$$

Equations 1.1 and 1.2 are only valid for homogeneous populations of dye molecules in quenching medium, such as solution-based systems, producing linear Stern–Volmer relationship of [O$_2$] versus τ^{-1} and allowing simple two-point calibration. However, many of the existing O$_2$-sensitive materials exhibit pronounced heterogeneity which results in nonlinear Stern–Volmer relationship [43]. This should be considered in the mechanistic description (physical model) and experimental calibration of the sensor. Calibration usually involves measurement of sensor signal (τ or I) at several known O_2 concentrations (standards) under constant temperature (25–37 °C for biological objects), and fitting these data points to determine the function $[O_2] = f(\tau, I)$. Under equilibrium, the concentration of O_2 in solution (and within the sensor material) is related to the partial pressure of O_2 in the gas phase, according to Henry's law. Sensor calibration in lifetime scale, i.e. $[O_2] = f(\tau)$, can be regarded as absolute, and indeed there are some commercial O_2 sensor systems which operate with factory calibration [35]. However, blind application of the available calibration on a different instrument, measurement set-up or sample type imposes a risk of generating inaccurate O_2 values. For proper operation of the sensor and accurate determination of O_2 concentration without significant instrumental errors and measurement artefacts, periodic re-calibrations or at least once-off independent calibrations should be considered.

Among the common sources of errors in O_2 measurement, is variation of sample/sensor temperature during the measurement. Since most of the O_2 sensors display strong temperature dependence of calibration, temperature drift or instability can skew the results of O_2 measurement. Singlet oxygen, a by-product of the quenching process and highly reactive but rather short-lived form of O_2, is another cause of concern. Produced by photosensitisation, singlet oxygen mostly returns back to the ground state O_2. But it can also react with nearby molecules (dye, lipids, proteins, nucleic acids) and damage the sensor or biological sample and affect the measurements [44].

1.3 Phosphorescent Metalloporphyrins and O_2-Sensitive Materials on Their Basis

Within the group of luminescent dyes efficiently quenched by O_2 are Pt(II)- and Pd(II)-porphyrins [37, 45, 46]. These molecular structures exhibit phosphorescence lifetimes in the range of 20–100 µs for Pt-porphyrins and 400–1,000 µs for Pd-porphyrins, which provide them from moderate to high quenchability by O_2. They have intense absorption bands in the near-UV (370–410 nm, Soret band) and visible (500–550 nm, Q-bands) regions, bright, well-resolved emission (630–700 nm) with relatively high quantum yields at physiological temperatures in aqueous solutions and solid-state formulations [37]. Some of the related structures, namely, the Pt- and Pd-complexes of benzoporphyrins, porphyrin-ketones and azaporphyrins, have longwave-shifted absorption (Q-bands > 600 nm) and phosphorescence in the very-near infrared region (700–900 nm). They are better suited for in vivo applications, but less compatible with standard photodetectors, such as PMT tubes. The structures of some dyes employed in O_2 sensors are shown in Fig. 1.1.

Phosphorescent O_2-sensitive materials are designed to attain the required physical, chemical, biological and O_2 quenching properties. For optimal analytical performance, sensor chemistry needs to be tailored to specific application, detection

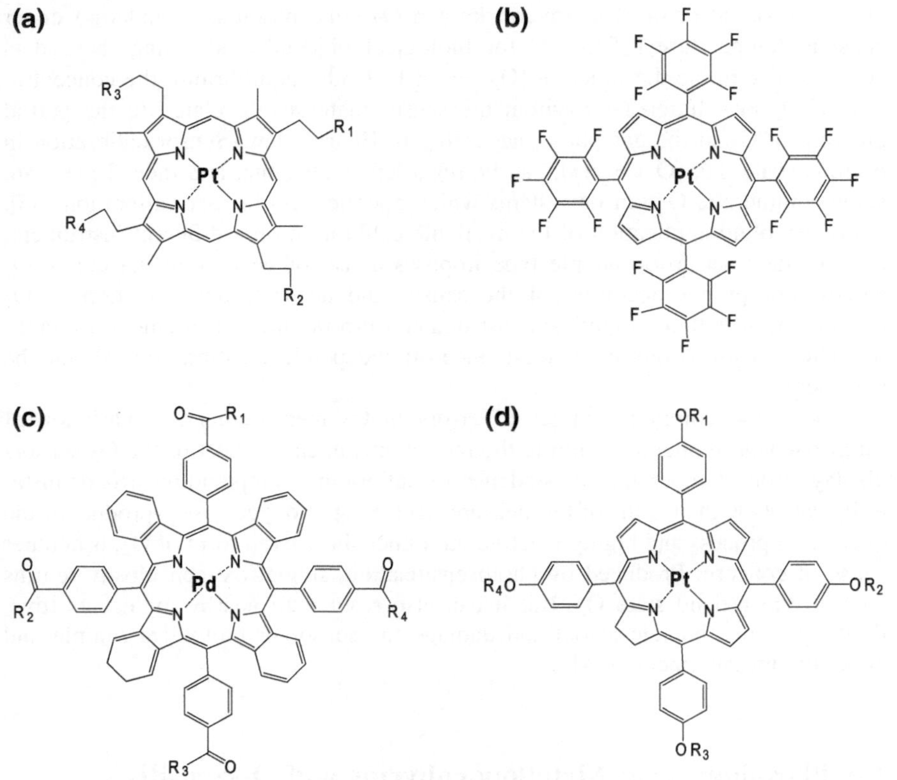

Fig. 1.1 Indicator dyes of porphyrin origin commonly used in O$_2$-sensitive materials. The derivatives of **a** Pt(II)-coproporphyrin–I (PtCP), R1 = R2 = R3 = R4 = COOH; **b** Pt(II)-meso-tetra-pentafluorophenyl porphyrin (PtPFPP); **c** meso-tetra(4-carboxyphenyl)tetrabenzoporphyrin (TPCTBP), R1 = R2 = R3 = R4 = OH; **d** meso-tetraarylporphyrin, R1 = R2 = R3 = R4— dendrimeric residues

platform and biological object being used. Thus, sensor excitation and emission spectra and photophysical properties (brightness, photostability) can be tuned by changing the macrocycle (e.g. CP, PFPP and TPCTBP). O$_2$ quenching efficiency and measurement range can be tuned by changing the central metal ion or microenvironment of the dye. Pt-porphyrins are less quenched and therefore better suited for the ambient O$_2$ range (0–200 µM), while Pd-porphyrins—for the low range 0–50 µM O$_2$. By introducing a dendrimeric shell or changing the polymeric matrix for dye encapsulation, one can alter the sensitivity to O$_2$ quenching [46–48]. Hydrophilicity can be improved by choosing the derivatives with polar side substituents on the macrocycle (e.g. CP, TPCTBP dendrimers), or by conjugating parent dye to a hydrophilic macromolecular carrier.

O$_2$ sensor material can be prepared as a macroscopic solid-state coating, microsensor deposited on the tip of optical fibre or liquid formulation—probe. Solid

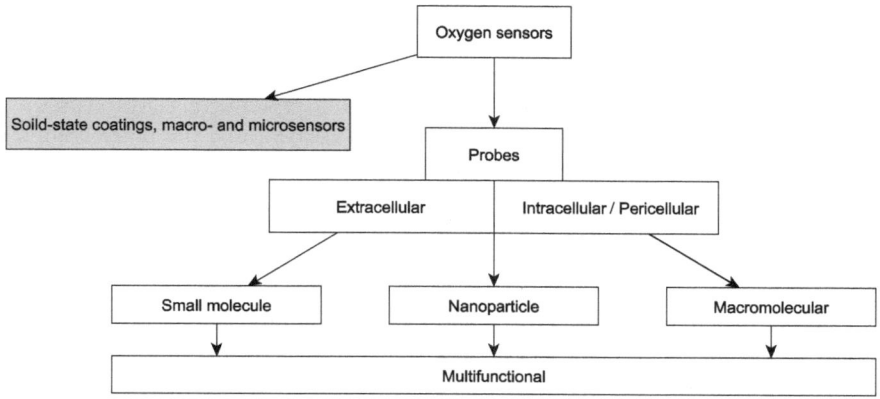

Fig. 1.2 Classification of O$_2$-sensing materials

state sensors have been used in coated microwell plates (BD Biosensor plate [49]), for the analysis of microbial and cellular respiration on a fluorescent reader (e.g. Mocon-Luxcel GreenLight® and Seahorse XF systems). O$_2$ microsensors [50] were applied to probe O$_2$ gradients in heterogeneous samples for single cell analysis [51], in microfluidic devices [52]. A number of solid-state O$_2$ sensors for O$_2$ measurement are produced commercially, for example, by Presens, Oxysense, Oxford Optronics, Mocon, Pyro Science, however, their main limitation is the lack of flexibility. Recent progress in this area is reviewed extensively [34, 52], so these systems are outside the scope of this book.

On the other hand, soluble O$_2$ probes provide greater flexibility and convenience for the users, and an extended range of applications [32]. The sensor can be added to the sample, dispensed, injected into tissue or animal and its working concentration is adjustable. Such a probe can be realised as a small molecule or supramolecular probes, nanoparticle and microparticle structures, which can also be combined with additional chemical, photophysical or biological functionality. Phosphorescent O$_2$-sensitive particles with magnetic properties have been described, which can be precipitated or localised within a sample with a magnet [48]. The main types of O$_2$ sensor materials are shown in Fig. 1.2.

1.4 Phosphorescent Probes for Sensing Cellular O$_2$

Within the group of *soluble O$_2$ probes* several categories can be defined. The small molecule probes are based on hydrophilic phosphorescent dyes or their derivatives bearing multiple polar or charged groups which provide them solubility in aqueous media. However, such probes have a tendency to bind non-specifically to proteins, cells and surfaces, display heterogeneity of their O$_2$ sensing properties and sensitivity to sample composition (pH, ionic strength, protein content).

These drawbacks can be partly addressed in the supramolecular probes in which several distinct functionalities are assembled together with the phosphorescent moiety in one chemical entity. Examples include the conjugates of PtCP dye with hydrophilic macromolecular carriers such as PEG or proteins (e.g. MitoXpress^{TM} probe [53] used in cell based in vitro screening assays), and the more complex dendrimeric probes developed for O$_2$ imaging in tissue and vasculature [46]. In such dendrimeric probe (see Fig. 1.1), four peripheral carboxylic groups of the *meso*-substituted Pd/Pt-(benzo)porphyrin are modified with dendritic polyglutamic chains that shield the phosphor and reduce the influence of pH, ionic strength and medium components on the dye emission and quenching by O$_2$.

In order to achieve controlled and specific localisation of the sensor within the biological sample or bring additional functionality such as targeting the probe to extracellular, intravascular, intracellular or pericellular localisation, sensor material can undergo further chemical modification or coupling with a suitable delivery vector. Thus, to make the dendrimeric probe more soluble in aqueous media (plasma), prevent penetration inside the cells and keep it in the bloodstream, an additional hydrophilic shell was introduced by PEGylation [47]. To deliver O$_2$ probe inside the cell or to the cell surface, supramolecular structures are produced comprising the conjugates of phosphorescent dyes with cell-penetrating, intracellular targeting peptide sequences or receptor molecules (e.g. lectins or antibodies) [39, 54–57].

Nanoparticle-based probes undergo active development [58–62]. These structures typically have a size of 30–200 nm and consist of a polymeric matrix in which the indicator dye(s) is/are incorporated by chemical linkage to the polymer backbone or surface groups, or by physical inclusion in a gel, co-precipitation and formation of core–shell structures [61]. Various fabrication methods allow flexibility in the choice of indicator dyes, nanosensor matrix, size and surface modification. Thus, hydrophobic dyes, structures lacking functional groups (i.e. not suitable for the other probe types) and pairs of dyes (ratiometric or FRET-based O$_2$ sensing) can be introduced in such systems [62, 63]. A number of biocompatible polymers and co-polymers have been used, including polyacrylamide, silica, polystyrene, polyfluorene and hydrogels. Other advantages of the nanoparticle O$_2$ probes are the possibility to achieve high specific brightness and photostability, relative ease of fabrication and tuning of sensor properties. The challenges are: larger size compared to the molecular probes, variable size, distribution and physical properties, difficulties in controlling the composition and structure during fabrication, instability under prolonged storage (drying and sterilisation can be problematic), toxicity and lack of biocompatibility in in vivo applications for many of such probes.

Initially, cell-impermeable O$_2$ probes were applied to monitor bulk oxygenation and OCR of respiring samples. This approach has been productive, with a number of applications and screening systems developed and adopted by many users (see Chap. 2). Nowadays, there is a growing demand in probes that have different and better defined location within the biological sample, particularly in cells, tissue, organs and whole organism. There is also a growing demand in measurement techniques and systems that allow probing and imaging of different O$_2$ pools in microscopic and macroscopic biological objects with high spatial resolution. With the advancement of probe and material chemistry and optical instrumentation, particularly fluorescence-based live cell imaging systems and sensitive time-resolved fluorescent readers, localised and targeted O$_2$ sensing approaches have become more common and available for ordinary users. Extensive experience of targeting small molecules to the cells and tissues (e.g. tumours) in drug delivery, MRI imaging and cancer therapy have been taken on board in O$_2$ sensing with a number of different types of probes with cell-penetrating capability, targeted to the membrane of mammalian cells and intracellular compartments described recently. These probes have opened the possibility to measure intracellular and pericellular O$_2$ concentrations and O$_2$ gradients between different compartments of respiring samples and within mammalian cells [55, 64].

The family of cell-targeted and intracellular probes have enabled in situ measurement of O$_2$ directly inside the cell, at cell surface and potentially in the mitochondria where most of O$_2$ gets consumed in mammalian cells and in peroxisomes for macrophages. In conjunction with high-resolution live cell imaging technique, this strategy provides the possibility to study intracellular O$_2$ gradient(s) with high selectivity, sensitivity and spatial resolution. This gives researchers a new level of detail about mitochondrial function, cell bioenergetics and biological roles of O$_2$. Rapid advancement of O$_2$-sensitive materials, new ways of intracellular delivery of small and large molecules and nanoparticle structures (e.g. by protein transduction domains, cell-penetrating peptide vectors) further extends our capabilities and allows new applications and O$_2$ sensing schemes.

The distinct photophysical characteristics of the phosphorescent O$_2$ probes provide large scope for multiplexing with other probes and parameters of cellular function, including Ca^{2+}, cellular ATP, NADH, mitochondrial and plasma membrane potentials, protein markers and fluorescent tags (GFP family). Several O$_2$ probes can be used with the same sample to monitor O$_2$ levels in different cellular compartments (intracellular, pericellular and extracellular O$_2$) and their dynamics upon changing cellular environment.

Some common phosphorescence-based O$_2$-sensitive probes designed for biological applications and their main photophysical and operational characteristics are described in Table 1.1.

1.5 Detection Modalities

Quenched-phosphorescence O$_2$ sensing can be realised by simple intensity-based, ratiometric or lifetime measurements. The main detection modalities are shown in Fig. 1.3.

Measurement of probe phosphorescence intensity is useful for qualitative and semi-quantitative assessment. The intensity signal is inversely related to O$_2$ concentration [see Eq. (1.2)]. However, in this mode, O$_2$ calibration is rather unstable

Table 1.1 An overview of O$_2$ sensing probes tested in biological applications (modified from [32])

Probe name and type	Application probe location	Equipment, detection mode	K_{s-v}, $\mu M^{-1}/\tau_0$, μs	Status, comments	References
Extracellular probes					
PdTPCPP (conjugated to BSA)—SM	O$_2$ mapping in tissues *Probe in blood/ vasculature*	Ex—416, 523 nm Em—690 nm Scanning phosphorescence quenching microscopy	0.382/ ~700	Quantitative Point measurement Used by several labs	[31, 65–70]
Oxyphor R2 (PdTPCPP dendrimer)—SM	O$_2$ mapping in tissues *Probe in blood/ vasculature*	Phase fluorometry: Ex—524 nm Em—690 nm 4 mm light guide; Two-photon microscope	0.343/640 (38 °C, pH 7.4)	Quantitative Point measurement Used by several labs	[46, 71]
Oxyphor G2 (PdTCPTBP dendrimer)—SM	Tissue O$_2$ gradients in vivo *Probe in blood/ vasculature*	Wide field FLIM: Ex—440/632 nm, Em—790 nm	0.086/251 (38 °C, pH 7.4)	Quantitative Used by several labs and with different models	[46, 72]
MitoXpress (PtCP conjugate)—SM	OCR by cells, mitochondria, enzymes. Assessment of cell bioenergetics *Probe added to the medium*	TR-F reader, RLD: Ex—340–420 nm Em—640–660 nm	0.04/67	Quantitative. Used by many labs. Validated in drug and toxicity screening	[53, 55, 73–75]
PtP-C343 (PtTAOP-Coumarin 343 dendrimer)—SM	Tissue O$_2$. In vivo O$_2$ gradients *Probe in blood/ vasculature*	Two-photon FLIM: Ex—840 nm Em—682 nm	>0.11/60	Quantitative. Require special equipment and setup Used in several labs	[47, 76, 77].

(continued)

Table 1.1 (continued)

Probe name and type	Application/probe location	Equipment, detection mode	K_{s-v}, μM^{-1}/$τ_0$, μs	Status, comments	References
Oxyphors R4 and G4	Tissue O$_2$ gradients in vivo. Tumour imaging *EC probe in blood/ vasculature/ interstitial space*	Wide-field FLIM Ex—428, 530 nm (R4); 448, 637 (G4) Em—698 (R4), 813 (G4)	R4: 0.098/681 (37 °C, pH 7.2) G4: 0.083/218 (38.2 °C, pH 7.2)	Quantitative. Require special equipment and setup	[78]
PS-NP (polystyrene NP doped with PdTPTBP and DY635—reference)	Targeted tumour in vivo imaging *EC probe*	Ratiometric-based Lifetime-based Ex—635 nm Em—670 nm (reference); 800 nm (O$_2$ sensitive)	ND	Quantitative. May be used for intracellular measurement with modified coating	[79]
Pericellular probes					
ER9Q-PtCP (PtCP protein conjugate)—SM	Cell oxygenation assessment of intracellular O$_2$ gradient *Stains plasma membrane of cultured cells*	TR-F reader, RLD: Ex—340–420 nm Em—640–660 nm	0.046/55	Quantitative 2 cell lines tested (PC12, MEF)	[55]
Intracellular probes					
MitoXpress (PtCP conjugate)—SM	Cell oxygenation. Metabolic responses *Impermeable probe delivered into the cell with Endo-Porter*	TR-F reader, RLD: Ex—340–420 nm Em—640–660 nm	~0.04/67	Quantitative. Facilitated loading required (24–28 h); 6 cell types tested (cell-specific)	[64, 80–83]

(continued)

Table 1.1 (continued)

Probe name and type	Application probe location	Equipment, detection mode	K_{s-v}, $\mu M^{-1}/\tau_0$, μs	Status, comments	References
O$_2$ PEBBLEs (PtOEPK & OEP, ormosil)—NP	Cell oxygenation *Impermeable probe delivered into the cell with gene gun*	Ratiometric wide field imaging Ex—568 nm, Em—620/750 nm	0.032/ND	Semi-quantitative (relative) Stressful loading 1 cell type tested	[84]
Ru(II)-(py)$_3$-R$_8$ (peptide conjugate)—SM	O2 mapping in cells *Cell-permeable, self-loading probe*	Wide field FLIM: Ex—460 nm Em—607 nm	ND	Semi-quantitative 1 cell type tested	[39]
Cell penetrating PtCP peptide conjugates: PtCPTE-CFR$_9$, PEPP0-5, T1-T4—SM	Cell oxygenation, Metabolic responses *Cell-permeable, self-loading probe*	TR-F reader, RLD: Ex—340–420 nm Em—640–660 nm Intravital confocal imaging was also demonstrated	~0.006/70	Quantitative >6 cell lines tested Controlled sub-cellular location	[54, 56, 85, 86]
PtOEP/PDHF and PFO—NP	O$_2$ mapping in cells *Probe uptake by macrophages*	Ratiometric wide field imaging: Ex—350 nm Em—440/650 nm	ND	Semi-quantitative One cell line tested (macrophages) particle variability, require UV excitation	[62]
Near infrared PAA NPs (Oxyphor G2 probe in PAA gel, with peptide coat)—NP	Cell oxygenation *Cell permeable, self-loading probe*	Wide field and confocal imaging: Ex—633 nm Em—790 nm	0.034/ND (37 °C) (without cells)	Quantitative Several cell lines tested. High probe concentrations used Cross-sensitivity to pH	[59]

(continued)

Table 1.1 (continued)

Probe name and type	Application probe location	Equipment, detection mode	K_{s-v}, μM^{-1}/τ_0, μs	Status, comments	References
RGB NPs (PtPFPP & BCPN in aminated polystyrene)—NP	Cell oxygenation *Cell permeable, self-loading probe*	Wide-field RGB imaging: Ex—330–380 nm Em—Red, Green	~0.0083/NA	Semi-quantitative 1 cell line tested. NP variability, long loading—48 h	[63]
MitoXpress-Intra (NANO2, PtPFPP in RL100 polymer)—NP	Cell oxygenation. Metabolic responses, cell bioenergetics *Cell permeable, self-loading probe*	TR-F reader: RLD Ex—340–420 nm Em—640–660 nm wide-field FLIM and confocal O$_2$ imaging	0.04/69	Quantitative >5 cell lines tested High brightness and photostability	[55, 87]
Cell penetrating IrOEP peptide conjugates: Ir1, Ir2,—SM	Cell oxygenation, Metabolic responses *Cell-permeable, self-loading probe*	TR-F reader, RLD: Ex—390 nm Em—650 nm	Ir1: 0.074(Ksv1)/58 Ir2: ND/69	Quantitative >5 cell lines tested	[57]
MM2 (PtPFPP, PFO in RL100 polymer)—NP	Cell oxygenation Metabolic responses, cell bioenergetics *Cell permeable, self-loading probe*	TR-F reader: RLD, FLIM, ratiometric intensity, multiphoton microscopy Ex—400 nm (single photon), 760 nm (two photon) Em—430, 650	0.04/61	Quantitative Tested with 2D and 3D cell samples	[88]

Note Ks–v constants were calculated based on the published data. At normal atmospheric pressure O$_2$ has 160 mmHg with dissolved concentration ~200 μM or 4,950 ppm

PAA polyacrylamide hydrogel; *PDHF* poly(9,9-dihexylfluorene); *PFO* (poly(9,9-diheptylfluorene-alt-9,9-di-p-tolyl-9H-fluorene)); *SM* supramolecular; *NA* not applicable; *ND* no data reported; *NP* nanoparticle-based; *BCPN* butyl-N-(5-carboxypentyl)-4-piperidino-1,8-naphthalimide

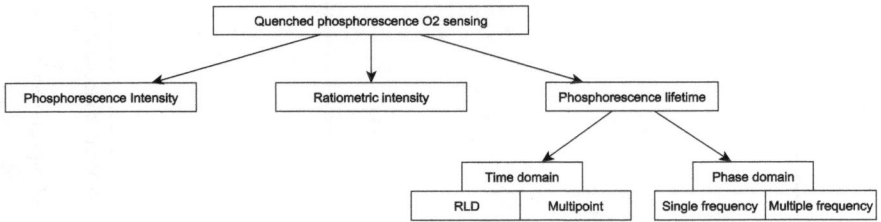

Fig. 1.3 Measurement modalities in O$_2$ sensing

and influenced by probe concentration, its photodegradation, measurement geometry, optical properties of the sample, drift and noise of the detector and light source. On the other hand, under standard conditions, relative changes in O$_2$ and OCR (e.g. treated versus untreated cells) can be measured easily and reliably in this mode.

Introduction of a reference (O$_2$-insensitive) dye in the sensor allows parallel intensity measurements in two spectral channels and determination of O$_2$ concentration from the ratio of the O$_2$ sensitive and O$_2$-insensitive signals. This approach overcomes many of the limitations of single channel measurements and makes O$_2$ calibration more stable and suitable for quantitative O$_2$ sensing. In certain cases, linear relationship of the ratio signal and O$_2$ concentration can be achieved. Nonetheless, the signal ratio can be influenced by the sample and detection system, especially when the two specific signals are of moderate or low intensity (low signal to noise/blank ratio). Factors such as instability of the detector and excitation source, differential scattering, autofluorescence and detector dark counts in the two spectral regions, different photobleaching rates of the two dyes can contribute and cause a drift or shift in O$_2$ calibration. Due to the relatively simple optical set-up, well-established instrumentation and measurement procedures for fluorescence ratiometric and intensity-based O$_2$ sensing can be implemented on a wide range of detection platforms available in ordinary research labs. However, one should use them with care to avoid measurement artefacts and experimental errors in O$_2$ quantification.

Phosphorescence lifetime-based O$_2$ sensing is by far regarded as the most stable and accurate modality for O$_2$ sensing [33]. Although instrumentation for luminescence lifetime measurements is somewhat more complex and less common than intensity-based systems, it is rapidly gaining popularity in life sciences since it can overcome many of the limitations of fluorescence intensity-based systems and provide higher confidence and stability in O$_2$ measurement. Measurement of phosphorescence lifetimes, which lay in the microsecond time domain, is technically simple, but often requires dedicated hardware, signal acquisition and processing algorithms to implement reliable and on-the-fly determination of probe lifetime. One such method is phase-fluorometry, in which the sample is excited with periodically modulated light (sine of square wave excitation at kHz frequencies) and detector optoelectronics measures the phase shift of luminescent signal with respect to excitation, ϕ. From the measured ϕ (degrees angle) phosphorescence lifetime of the probe is calculated as:

$$\tau = \text{tg}(\phi)/2\pi v \tag{1.3}$$

where v is modulation frequency of excitation (Hz), and O_2 concentration is calculated according to Eq. (1.2) [89].

An alternative method is based on direct measurement of the phosphorescence decay under short-pulse excitation (<10 μs), using a multi-channel scaler with photon counting detector or Time-Correlated Single Photon Counting (TCSPC) board [42]. A simplified version of this method is called Rapid Lifetime Determination (RLD), in which emission intensity signals (F_1, F_2) are collected at two different delay times (t_1, t_2) after the excitation pulse and lifetime is calculated as follows [82]:

$$\tau = (t_2 - t_1)/\ln(F_1/F_2) \tag{1.4}$$

Time-resolved detection in the microsecond range also allows effective elimination of sample autofluorescence and scattering, providing large improvement in sensitivity and selectivity of probe detection and reduced interferences. As a result, RLD usually provides good accuracy and resolution in the measurement of phosphorescence lifetimes of O_2 sensitive materials (including the intracellular O_2 probes) and quantification of O_2 concentration in complex biological samples. It should be noted though that RLD operates reliably only when blank signals are low and specific signals high, i.e. S:N > 5 [82]. Modern instruments often have built-in microsecond time-resolved fluorometry (TR-F) and lifetime measurement capabilities, which make them suitable for O_2 sensing with the phosphorescent porphyrin probes. Examples include the multi-label fluorescence readers for assays in microplates, screening systems, wide-field or laser-scanning microscopes that support time-domain and phase-domain FLIM mode. At the same time, such instruments need to be assessed thoroughly to ensure their performance with the probe, including the sensitivity, reproducibility and accuracy of lifetime measurements, as well as temperature control and uniformity of readings across the plate. Many commercial systems cannot provide adequate performance in the measurement of short lifetimes of Pt-porphyrins and/or longwave emission of meso-substituted and benzo-porphyrins.

The core detection modalities described above can be realised as 'cuvette' format, in which the optical signal and O_2 concentration are measured at one point or for the whole sample (macroscopic) on a suitable luminescent spectrometer or reader. Using epifluorescence alignment of the optics and an X–Y stage, measurement of multiple points/samples or two-dimensional (2-D) scanning with sub-mm resolution can be implemented in multi-well plates or other substrates. If an O_2 probe is introduced in a particular compartment (e.g. intracellular or pericellular), one can achieve measurement of *local* concentrations and gradients of O_2 within the sample. Alternatively, quenched-phosphorescence detection can be coupled with an imaging detector, thus enabling O_2 imaging within the sample, for example, on a fluorescence

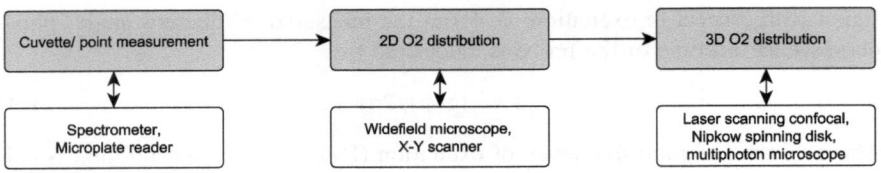

Fig. 1.4 O_2 measurement tasks and detection platforms for various experimental needs

microscope, live cell imaging (LCI) or macroscopic imaging systems. The main measurement tasks and detection options are shown schematically in Fig. 1.4).

Relatively simple and inexpensive wide-field fluorescence microscopes operating in intensity and ratiometric mode can provide 2-D visualisation of respiring objects stained with an O_2 probe, with sub-cellular spatial resolution. Relative oxygenation and changes related to sample respiration activity can be monitored performing time-lapse imaging experiments. Furthermore, confocal microscopy represented by laser-scanning and spinning disk fluorescence LCI systems enables the analysis of 3-D O_2 distribution in complex objects with sub-micron spatial resolution. Compared to the traditional microscopes exciting the luminophores in one-photon mode, the multi-photon imaging systems employ high-power femtosecond NIR lasers for excitation which provide better contrast and spatial resolution and deep penetration into tissue (>500 µm). However, such systems are more expensive and they require special indicator dyes with large cross-section of two-photon absorption, and to be able to see the long-decay emission of the probes and their response to O_2, system hardware and software need to be specially tuned [90, 91]. A number of dedicated O_2 probes with two-photon and FRET antennae, imaging systems and applications on their basis have been described recently, and this area continues to develop rapidly [47, 62, 78, 92].

Fluorescence/Phosphorescence Lifetime Imaging Microscopy (FLIM/PLIM, also called PQM—Phosphorescence Quenching Microscopy [93]) enables more reliable visualisation of O_2 distribution in complex biological samples, and accurate quantification of O_2. Wide-field microscopes equipped with gated CCD camera and LED/laser excitation providing trains of ns-µs pulses at kHz frequency can generate 2-D O_2 images with single cell resolution [66, 67, 76, 94]. Following each excitation pulse and a time delay (variable), emitted photons are collected by the camera over the measurement window time and integrated over a number of pulses to generate an intensity frame. This is repeated at several delay times, and from this set of frames emission decay is reconstructed and lifetime is determined for each pixel of the CCD matrix. By applying probe calibration function (determined in a separate experiment), lifetime images of the sample can be converted into O_2 concentration map.

For the laser-scanning systems, emission lifetimes are measured sequentially for each pixel with a PMT or photodiode detector, processed by the software to generate 2-D images of Z-stacks which are then assembled together to produce 3-D O_2 maps. Several custom-built PLIM systems employing detection and lifetime determination with TCSPC under both one-photon and two-photon excitation have been described in recent years and applied to O_2 imaging of live tissue in animal models. A dedicated hybrid ns/ms FLIM/PLIM hardware optimised for phosphorescence lifetime imaging and O_2 sensing experiments is produced commercially, by Becker & Hickl GmbH for example [42, 95].

In high-resolution microscopic imaging of O_2, samples are exposed to high illumination intensities and probe photostability becomes a critical issue. Many O_2–sensitive dyes and probes on their basis lack photostability. Perfluorinated PtPFPP dye is regarded as one of the most photostable dyes for such applications [87]. O_2 imaging experiments require thorough optimisation to produce sufficiently high, reliably measurable luminescent signals along with low phototoxicity, cell damage and photobleaching. In addition, careful calibration (measuring probe signal at several known pO_2 levels) is required to be able to convert raw fluorescence intensity images into O_2 concentration maps.

1.6 Measurement Formats Used in Optical O_2 Sensing

The various probe chemistries, detection modalities and platforms enable realisation of O_2 sensing in different measurement formats, thus making it versatile and suitable for a broad range of analytical tasks and applications. Some of these formats allow high sample throughput, high information content and multiplexing with the other biomarkers and parameters of cellular function.

The traditional set-up for O_2 respirometry is an air-tight cell, such as quartz cuvette with a stopper, which accommodates the biological sample along with the probe and is measured on a spectrometer to determine probe phosphorescent signal and changes over time and relate them to O_2 concentration or OCR, respectively. For accurate quantification, the anaerobic cuvette should contain no headspace or bubbles (air has much higher capacity for O_2 than aqueous media, and this may skew the results), be sealed, maintained at constant temperature (37 °C is optimal for eukaryotic cells), and stirred to distribute the respiring matter uniformly. At the same time, there is a growing need to conduct rapid, parallel O_2 sensing experiments with large number of biological samples (different cells, drugs, conditions, replicates and controls), to use the existing detection and screening platforms and miniaturise the bioassays. Sets of several anaerobic micro-cuvettes can be aligned on a multi-cell holder of a fluorescent reader, but this still does not provide the required sample throughput and requires modifications to the conventional anaerobic cuvette format. Examples of specialised substrates for optical O_2 sensing and respirometry include narrow-bore capillary cuvettes from the LightCycler® system measured on a carousel by dedicated detector (originally developed for quantitative PCR), standard microtiter plates with and without oil seal, low-volume sealable microplates, microfluidic biochips and perfusion chambers [36, 73, 75, 96, 97]. The common measurement formats are shown schematically in Fig. 1.5.

In particular, conventional microtiter plates provide large time savings, and reduced use of valuable and perishable biomaterials with drifting activity. They

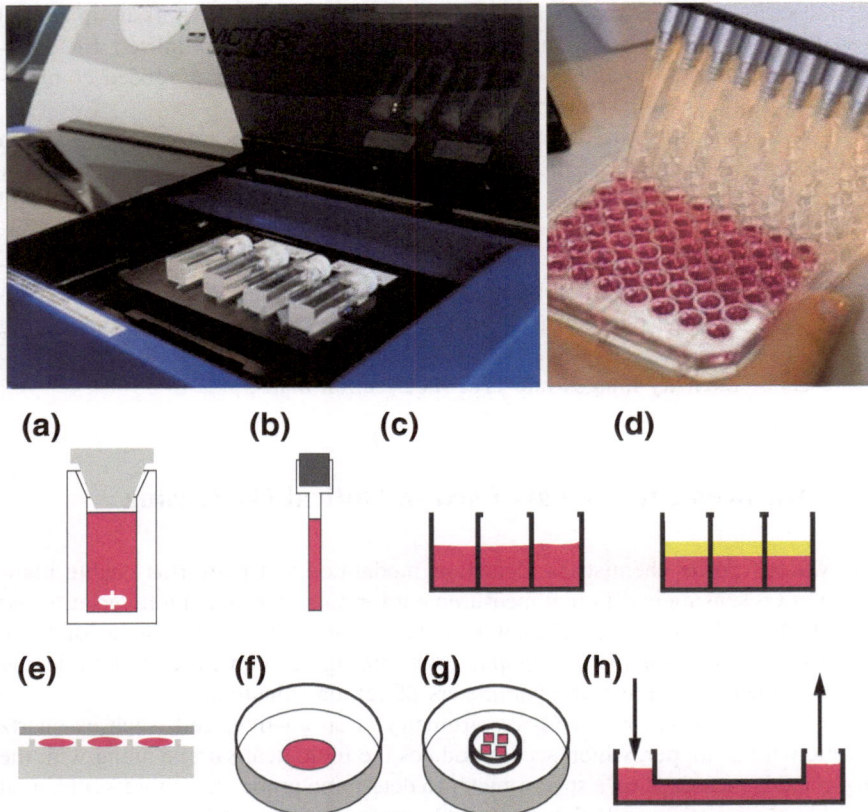

Fig. 1.5 Different measurement set-ups and substrates employed in optical O$_2$ sensing. *Top panel* Anaerobic micro-cuvettes (*left*) and 96-well plate with mammalian cells (*right*) being prepared for respirometric measurements on a commercial TR-F reader. *Bottom panel.* **a** sealed quartz cuvette with stirrer; **b** glass capillary cuvette (operate on the LightCycler® quantitative PCR instrument **c, d** microplate with sample wells unsealed (**c**) or sealed with the layer of oil (**d**); **e** Sealable low-volume microplate (Luxcel); **f, g** glass-bottom minidish for biological samples adapted for analysis of single (**f**) or multiple (**g**) samples (Ibidi); **h** Microfluidic biochips and perfusion flow chambers. The biomaterial analysed is shown in *pink colour*

facilitate assay miniaturisation (96- and 384-well plates are the most common), the use of automated liquid handling equipment (multichannel pipettes, dispensers and robots) and multi-label fluorescent or TR-F readers available in many labs. On the other hand, respirometric assays in microplates often have compromised performance. Thus, due to partial sealing of samples (oil seal and plastic body of the plate still allow ambient O$_2$ to diffuse in), assay sensitivity is reduced and usually *relative* but not absolute OCRs can be assessed reliably. Slow temperature equilibration requires care when preparing the plate, analysing signal profiles (negative slopes at the start of the assay are common) and getting consistent results in all wells across the plate.

Another measurement format represents a vessel with medium and respiring material exposed to a gaseous atmosphere such as ambient air. This format can be used to detect microbial growth/respiration in microplates with built-in O$_2$ sensors (BD Biosciences). Generally, such assays require relatively high levels of respiration and are more easily affected by sample distortion (respiration profiles are less reproducible compared e.g. to oil-sealed samples). With the development of intracellular (cell-permeable) O$_2$ probes, the range of analytical tasks that can be conducted in this manner have been extended. In particular monitoring of in situ oxygenation, respiratory activity and responses to metabolic stimulation of adherent cell cultures [56, 64, 82], analysis and imaging of O$_2$ in complex objects such as heterogeneous populations of cells, tissue slices, spheroids, small organisms, experiments under hypoxia in which the operator can precisely control atmospheric pO$_2$ and/or cellular O$_2$ levels, and conduct mechanistic biological studies under such conditions can be conducted using this simple format. It is also quite common in imaging experiments where glass-bottom minidishes with cells are commonly used. The latter can also be used with silicon microchamber inserts dividing the sample into several compartments (Fig. 1.5).

1.7 Biological Applications of Optical O$_2$ Sensing

The existing range of different probes, measurement formats and detection modalities for O$_2$ sensing open a large scope for the use of these techniques in various biological and physiological studies. One of the main advantages of the optical O$_2$ sensing technique is the possibility of contactless and minimally invasive measurements with gentle biological samples. The probe can be simply introduced into the sample and then interrogated with an external detector which measures probe luminescent signal and converts it into O$_2$ concentration.

These applications can be grouped into several categories:

- *Analysis of homogeneous, macroscopic samples* for example, monitoring of enzymatic O$_2$ consumption, quantification of activity and inhibition of important enzymes (e.g. cyclooxygenase, monoamino oxidase or cytochrome p450 oxidase [97]), determination of their substrates and metabolites present in test sample, enzyme biosensors (e.g. for glucose, lactate). Similarly, OCR and activity of mitochondrial preparations (e.g. from rat liver, heart, brain, human tissue) can be analysed under different conditions (e.g. State 2 and State 3 respiration, inhibition and uncoupling with drugs [3]), so metabolic activity and proliferation rate of suspension eukaryotic cells can be assessed (yeast, mammalian cells [36, 53, 98]). Quickly proliferating microbial cells which produce characteristic respiration profiles with a steep transition from aerated to deoxygenated condition, have to be analysed differently [99]. All these respirometric assays and applications can be conducted in a simple mix-and-measure procedure in standard 96-or 384-well plates on a standard plate reader.

- *Physiological studies with cultures of adherent mammalian cells (monolayers, 2-D models)* is inarguably the largest and most important group of in vitro assays. They are now widely used in various areas of life sciences and biomedical research, including general cell biology, disease models, drug development, biochemical toxicology, drug safety assessment, environmental monitoring. It includes the analysis of cell respiration, assessment of cell bioenergetics and metabolic status (in conjunction with the other biomarkers), comparison of normal and transformed cells, monitoring changes in cellular function and relating them to disease state or therapeutic treatment. Such assays are easy to perform in microtiter plates with extracellular O$_2$ probes such as MitoXpress [35, 75, 81, 100–102]. OCR measurements can be multiplexed with other probes (extracellular acidification, cellular ATP, Ca^{2+}, ROS, MMP) to achieve high-throughput multi-parametric assessment.
- *Control of cell oxygenation and experiments under hypoxia.* Live cells constantly consume O$_2$, which provides them energy in the form of ATP and also acts as a substrate in numerous biochemical reactions, thus acting as O$_2$ sink [7, 103]. Despite the efficient supply by the blood vessels and vasculature and rapid diffusion of O$_2$ across the cells and tissue, cells and tissues deoxygenate their environment and function under reduced O$_2$ levels (compared to ambient 21 % of O$_2$). On the other hand, most of the cell culture work is still performed at ambient O$_2$ (21 % in the atmosphere) which is regarded as a hyperoxia at which the cells experience oxidative stress and may behave differently to the in vivo conditions. This is particularly important for research in area of cancer and stem cells which normally reside in hypoxic or anoxic niches [104]. Intracellular O$_2$ probes provide useful tools for in situ control of oxygenation of cell monolayers and individual cells under ambient and hypoxic conditions, and to study adaptive responses of cells to hypoxia, drug action, signalling and cells physiology, particularly for neuronal cells [80, 105].
- In vitro *analysis of heterogeneous 2D and 3D respiring objects* including mixed cultures of different cells (co-cultures), 3-D scaffolds and spheroids (e.g. neurospheres), samples of animal tissue (slices) cultured under static or perfused conditions. Such systems represent the native microenvironment of mammalian cells in vivo more closely, and therefore represent more relevant cell models.
- In vivo *imaging* of tissue O$_2$ is of high fundamental and practical importance. Measurement of actual oxygenation in live respiring tissue (e.g. brain or muscle), localised O$_2$ gradients in the vasculature (blood vessels, capillaries) or tumour oxygenation can be realised using extracellular O$_2$ probes and phosphorescence lifetime-based O$_2$ imaging [65, 66, 68, 71, 72, 106–108]. This was also realised in plant cells [50]. In the last few years, wide-field FLIM systems and high-resolution confocal and two-photon laser-scanning systems [109] for imaging tissue O$_2$ were successfully used in complex in vivo and ex vivo studies. Thus, the dendrimeric probe PtP-C343 was injected in the blood stream and used to measure local oxygenation in rodent brain at different distances from arterial regions on a two-photon FLIM LCI system [76]. Such studies normally require special setup, measurement equipment, skills and ethical

approval for animal work. Due to our limited experience, we are not describing them in great detail.

- Ex vivo *imaging of O_2 in perfused organs and tissues*. This is performed similarly to the above in vivo sensing of O_2. Thus, mapping of O_2 in rodent retina [69, 72], dynamics of O_2 in individual frog skeletal muscle fibres [110]; imaging of oxygenation of tumours [79, 111] and O_2 in microcirculation [65, 66] were reported. Oxygenation of carotid body explant was monitored with the intracellular O_2 probe and correlated to cellular Ca^{2+} levels [85].

- *Assessment of intracellular O_2 gradients*. This area still remains obscure. Robiolio et al. reported O_2 gradients in neuroblastoma cells [112], but then other groups failed to detect such gradients in vascular [77] and hepatic Hep3B cells [30]. Parallel measurement of mitochondrial O_2 with endogenous protoporphyrin IX delayed fluorescence and extracellular O_2 with Oxyphor G2 probe revealed marginally small gradients: ~ 2 Torr in resting and ~ 4 Torr in uncoupled neuroblastoma and fibroblast cells [26]. For intact rat liver even at low ambient O_2, mito O_2 still had high values [25]. Significant intra-tissue heterogeneity and possible icO_2 gradient were reported for heart tissue [113]. Introduction of new O_2 probes targeted to intracellular compartments and cell membrane will help to advance this field [55], and clarify possible inconsistencies and experimental artefacts from the above studies.

- *Other groups including O_2 measurement in photosynthetic systems, small organisms and microfluidic biochips*. Plants produce O_2 by photosynthesis during light phase, and consume O_2 during dark phase [34, 50, 114]. Aquatic and underground organisms (i.e. round worms *C.elegans, zebrafish Danio rerio, Daphnia*) also experience hypoxia in their habitat [115]. Studies of behaviour of these model animals under hypoxic conditions and in various physiological and toxicological studies are on the rise. In situ control of oxygenation in cultures of these organisms, within individual animals and their microenvironment is important from the biological and physiological points of view. Other attractive models include artificially engineered mammalian tissues and organs, microfluidic devices, cell and tissue-based biochips [116].

A number of representative examples covering the above applications with different biological models and studies are described in greater detail in the following two experimental Chapters. Chapter 2 is focused on plate reader analysis of macroscopic samples, including eukaryotic and prokaryotic cells, spheroids, enzymes, small organisms, comparison of different cells, drugs and treatments. Chapter 3 describes on O_2 imaging in individual cells, complex 3D objects with high level of detalisation and generation of 2-D and 3-D O_2 maps and time profiles of oxygenation.

Acknowledgments This work was supported by the Science Foundation Ireland, grant 07/IN.1/ B1804 and the Ministry of Education and Science of Russian Federation, State Contract No 14.740.11.0909.

References

1. Semenza GL (2007) Life with oxygen. Science 318(5847):62–64
2. Devor A, Sakadzic S, Srinivasan VJ, Yaseen MA, Nizar K, Saisan PA, Tian P, Dale AM, Vinogradov SA, Franceschini MA, Boas DA (2012) Frontiers in optical imaging of cerebral blood flow and metabolism. J Cereb Blood Flow Metab 32(7):1259–1276. doi:10.1038/jcbfm.2011.195
3. Brand MD, Nicholls DG (2011) Assessing mitochondrial dysfunction in cells. Biochem J 435(2):297–312. doi:10.1042/bj20110162
4. Wilson DF, Finikova OS, Lebedev AY, Apreleva S, Pastuszko A, Lee WMF, Vinogradov SA (2011) Measuring oxygen in living tissue: intravascular, interstitial, and "tissue" oxygen measurements. In: LaManna JC, Puchowicz MA, Xu K, Harrison DK, Bruley DF (eds) Advances in experimental medicine and biology, vol 701. Springer, US, pp 53–59. doi:10.1007/978-1-4419-7756-4_8
5. Semenza GL (2007) Oxygen-dependent regulation of mitochondrial respiration by hypoxia-inducible factor 1. Biochem J 405(1):1–9
6. Wilson DF (2008) Quantifying the role of oxygen pressure in tissue function. Am J Physiol Heart Circ Physiol 294(1):H11–H13. doi:01293.200710.1152/ajpheart.01293.2007
7. Erecinska M, Silver IA (2001) Tissue oxygen tension and brain sensitivity to hypoxia. Respir Physiol 128(3):263–276
8. Jezek P, Plecitá-Hlavatá L, Smolková K, Rossignol R (2010) Distinctions and similarities of cell bioenergetics and the role of mitochondria in hypoxia, cancer, and embryonic development. Int J Biochem Cell biol 42(5):604–622
9. Lin J, Handschin C, Spiegelman BM (2005) Metabolic control through the PGC-1 family of transcription coactivators. Cell Metab 1(6):361–370
10. Bartrons R, Caro J (2007) Hypoxia, glucose metabolism and the Warburg's effect. J Bioenerg Biomembr 39(3):223–229
11. De Filippis L, Delia D (2011) Hypoxia in the regulation of neural stem cells. Cell Mol Life Sci 68(17):2831–2844. doi:10.1007/s00018-011-0723-5
12. Takahashi E, Doi K (1998) Impact of diffusional oxygen transport on oxidative metabolism in the heart. Jpn J Physiol 48(4):243–252
13. Palm F, Nordquist L (2011) Renal oxidative stress, oxygenation, and hypertension. Am J Physiol Regul Integr Comp Physiol 301(5):R1229–R1241. doi:10.1152/ajpregu.00720.2010
14. Wagner PD (2012) Muscle intracellular oxygenation during exercise: optimization for oxygen transport, metabolism, and adaptive change. Eur J appl Physiol 112(1):1–8. doi:10.1007/s00421-011-1955-7
15. Clark LC, Wolf R, Granger D, Taylor Z (1953) Continuous recording of blood oxygen tensions by polarography. J Appl Physiol 6(3):189–193
16. Wu C–C, Luk H-N, Lin Y-TT, Yuan C-Y (2010) A Clark-type oxygen chip for in situ estimation of the respiratory activity of adhering cells. Talanta 81(1–2):228–234
17. Jekabsons MB, Nicholls DG (2004) In situ respiration and bioenergetic status of mitochondria in primary cerebellar granule neuronal cultures exposed continuously to glutamate. J Biol Chem 279(31):32989–33000. doi:10.1074/jbc.M401540200
18. Yadava N, Nicholls DG (2007) Spare respiratory capacity rather than oxidative stress regulates glutamate excitotoxicity after partial respiratory inhibition of mitochondrial complex I with rotenone. J Neurosci 27(27):7310–7317. doi:10.1523/jneurosci.0212-07.2007
19. Williams BB, Khan N, Zaki B, Hartford A, Ernstoff MS, Swartz HM (2010) Clinical electron paramagnetic resonance (EPR) oximetry using India ink. In: Takahashi E, Bruley DF (eds) Advances in experimental medicine and biology, vol 662. Springer US, pp 149–156. doi:10.1007/978-1-4419-1241-1_21
20. Liu Y, Villamena FA, Sun J, Wang TY, Zweier JL (2009) Esterified trityl radicals as intracellular oxygen probes. Free Radic Biol Med 46(7):876–883

21. Bobko AA, Dhimitruka I, Eubank TD, Marsh CB, Zweier JL, Khramtsov VV (2009) Trityl-based EPR probe with enhanced sensitivity to oxygen. Free Radic Biol Med 47(5):654–658

22. Halevy R, Shtirberg L, Shklyar M, Blank A (2010) Electron spin resonance micro-imaging of live species for oxygen mapping. J Vis Exp 42:e2122

23. Wittenberg JB (1970) Myoglobin-facilitated oxygen diffusion: role of myoglobin in oxygen entry into muscle. Physiol Rev 50(4):559–636

24. Foster KA, Galeffi F, Gerich FJ, Turner DA, Müller M (2006) Optical and pharmacological tools to investigate the role of mitochondria during oxidative stress and neurodegeneration. Prog Neurobiol 79(3):136–171

25. Mik EG, Johannes T, Zuurbier CJ, Heinen A, Houben-Weerts JHPM, Balestra GM, Stap J, Beek JF, Ince C (2008) In vivo mitochondrial oxygen tension measured by a delayed fluorescence lifetime technique. Biophys J 95(8):3977–3990

26. Mik EG, Stap J, Sinaasappel M, Beek JF, Aten JA, van Leeuwen TG, Ince C (2006) Mitochondrial PO2 measured by delayed fluorescence of endogenous protoporphyrin IX. Nat Meth 3(11):939–945

27. Ashkenazi S, Huang S-W, Horvath T, Koo Y-EL, Kopelman R (2008) Photoacoustic probing of fluorophore excited state lifetime with application to oxygen sensing. J Biomed Opt 13(3):034023–034024

28. Potzkei J, Kunze M, Drepper T, Gensch T, Jaeger K-E, Buechs J (2012) Real-time determination of intracellular oxygen in bacteria using a genetically encoded FRET-based biosensor. BMC Biol 10(1):28

29. Takahashi E, Sato M (2010) Intracellular diffusion of oxygen and hypoxic sensing: role of mitochondrial respiration new frontiers in respiratory control. In: Homma I, Fukuchi Y, Onimaru H (eds) Advances in experimental medicine and biology, vol 669. Springer, New York, pp 213–217. doi:10.1007/978-1-4419-5692-7_43

30. Takahashi E, Takano T, Nomura Y, Okano S, Nakajima O, Sato M (2006) In vivo oxygen imaging using green fluorescent protein. Am J Physiol Cell Physiol 291(4):C781–C787. doi:10.1152/ajpcell.00067.2006

31. Vanderkooi JM, Maniara G, Green TJ, Wilson DF (1987) An optical method for measurement of dioxygen concentration based upon quenching of phosphorescence. J Biol Chem 262(12):5476–5482

32. Dmitriev RI, Papkovsky DB (2012) Optical probes and techniques for O(2) measurement in live cells and tissue. Cell Mol Life Sci CMLS. doi:10.1007/s00018-011-0914-0

33. Papkovsky DB (2004) Methods in optical oxygen sensing: protocols and critical analyses. Methods Enzymol 381:715–735. doi:10.1016/S0076-6879(04)81046-2S0076687904810462

34. Ast C, Schmälzlin E, Löhmannsröben H-G, van Dongen JT (2012) Optical oxygen micro- and nanosensors for plant applications. Sensors 12(6):7015–7032

35. Gerencser AA, Neilson A, Choi SW, Edman U, Yadava N, Oh RJ, Ferrick DA, Nicholls DG, Brand MD (2009) Quantitative microplate-based respirometry with correction for oxygen diffusion. Anal Chem 81(16):6868–6878. doi:10.1021/ac900881z

36. Hynes J, Natoli E, Jr., Will Y (2009) Fluorescent pH and oxygen probes of the assessment of mitochondrial toxicity in isolated mitochondria and whole cells. Curr Protoc Toxicol Chapter 2 Unit 2 16. doi:10.1002/0471140856.tx0216s40

37. Papkovsky DB, O'Riordan TC (2005) Emerging applications of phosphorescent metalloporphyrins. J Fluoresc 15(4):569–584. doi:10.1007/s10895-005-2830-x

38. Yoshihara T, Yamaguchi Y, Hosaka M, Takeuchi T, Tobita S (2012) Ratiometric molecular sensor for monitoring oxygen levels in living cells. Angew Chemie Int Ed 51(17):4148–4151. doi:10.1002/anie.201107557

39. Neugebauer U, Pellegrin Y, Devocelle M, Forster RJ, Signac W, Moran N, Keyes TE (2008) Ruthenium polypyridyl peptide conjugates: membrane permeable probes for cellular imaging. Chem Commun (42):5307–5309

40. Koren K, Borisov SM, Saf R, Klimant I (2011) Strongly phosphorescent iridium(III)–porphyrins—new oxygen indicators with tuneable photophysical properties and functionalities. Eur J Inorgan Chem 2011(10):1531–1534. doi:10.1002/ejic.201100089

41. Stern O, M. Volmer (1919) The fading time of fluorescence. Physikalishe Zeitschrift 20:183–188

42. Becker W, Bergmann A, Biskup C (2007) Multispectral fluorescence lifetime imaging by TCSPC. Microsc Res Tech 70(5):403–409. doi:10.1002/jemt.20432

43. Carraway ER, Demas JN, DeGraff BA, Bacon JR (1991) Photophysics and photochemistry of oxygen sensors based on luminescent transition-metal complexes. Anal Chem 63(4):337–342. doi:10.1021/ac00004a007

44. Schweitzer C, Schmidt R (2003) Physical mechanisms of generation and deactivation of singlet oxygen. Chem Rev 103(5):1685–1758. doi:10.1021/cr010371d

45. Rumsey WL, Vanderkooi JM, Wilson DF (1988) Imaging of phosphorescence: a novel method for measuring oxygen distribution in perfused tissue. Science 241(4873):1649–1651. doi:10.1126/science.3420417

46. Dunphy I, Vinogradov SA, Wilson DF (2002) Oxyphor R2 and G2: phosphors for measuring oxygen by oxygen-dependent quenching of phosphorescence. Anal Biochem 310(2):191–198

47. Lebedev AY, Troxler T, Vinogradov SA (2008) Design of metalloporphyrin-based dendritic nanoprobes for two-photon microscopy of oxygen. J Porphyr Phthalocyanines 12(12):1261–1269. doi:10.1142/S1088424608000649

48. Mistlberger Gn, Koren K, Borisov SM, Klimant I (2010) Magnetically remote-controlled optical sensor spheres for monitoring oxygen or pH. Anal Chem 82(5):2124–2128. doi:10.1021/ac902393u

49. Wang W, Upshaw L, Strong DM, Robertson RP, Reems J (2005) Increased oxygen consumption rates in response to high glucose detected by a novel oxygen biosensor system in non-human primate and human islets. J Endocrinol 185(3):445–455. doi:10.1677/joe.1.06092

50. Schmälzlin E, van Dongen JT, Klimant I, Marmodée B, Steup M, Fisahn J, Geigenberger P, Löhmannsröben H-G (2005) An optical multifrequency phase-modulation method using microbeads for measuring intracellular oxygen concentrations in plants. Biophys J 89(2):1339–1345

51. Molter TW, McQuaide SC, Suchorolski MT, Strovas TJ, Burgess LW, Meldrum DR, Lidstrom ME (2009) A microwell array device capable of measuring single-cell oxygen consumption rates. Sens Actuators B Chem 135(2):678–686

52. Thomas PC, Raghavan SR, Forry SP (2011) Regulating oxygen levels in a microfluidic device. Anal Chem 83(22):8821–8824. doi:10.1021/ac202300g

53. Hynes J, Floyd S, Soini AE, O'Connor R, Papkovsky DB (2003) Fluorescence-based cell viability screening assays using water-soluble oxygen probes. J Biomol Screen 8(3):264–272. doi:10.1177/1087057103008003004

54. Dmitriev RI, Ropiak H, Ponomarev G, Yashunsky DV, Papkovsky DB (2011) Cell-penetrating conjugates of coproporphyrins with oligoarginine peptides: rational design and application to sensing of intracellular O₂. Bioconjug Chem. doi:10.1021/bc200324q

55. Dmitriev RI, Zhdanov AV, Jasionek G, Papkovsky DB (2012) Assessment of cellular oxygen gradients with a panel of phosphorescent oxygen-sensitive probes. Anal Chem 84(6):2930–2938. doi:10.1021/ac3000144

56. Dmitriev RI, Ropiak HM, Yashunsky DV, Ponomarev GV, Zhdanov AV, Papkovsky DB (2010) Bactenecin 7 peptide fragment as a tool for intracellular delivery of a phosphorescent oxygen sensor. FEBS J 277(22):4651–4661. doi:10.1111/j.1742-4658.2010.07872.x

57. Koren K, Dmitriev RI, Borisov SM, Papkovsky DB, Klimant I (2012) Complexes of IrIII-octaethylporphyrin with peptides as probes for sensing cellular O₂. ChemBioChem 13:1184–1190. doi:10.1002/cbic.201200083

58. Koo Lee Y-E, Smith R, Kopelman R (2009) Nanoparticle PEBBLE sensors in live cells and in vivo. Ann Rev Anal Chem 2(1):57–76. doi:10.1146/annurev.anchem.1.031207.112823

59. Koo Lee Y-E, Ulbrich EE, Kim G, Hah H, Strollo C, Fan W, Gurjar R, Koo S, Kopelman R (2010) Near infrared luminescent oxygen nanosensors with nanoparticle matrix tailored sensitivity. Anal Chem 82(20):8446–8455. doi:10.1021/ac1015358

60. Coogan MP, Court JB, Gray VL, Hayes AJ, Lloyd SH, Millet CO, Pope SJA, Lloyd D (2010) Probing intracellular oxygen by quenched phosphorescence lifetimes of nanoparticles containing polyacrylamide-embedded [Ru(dpp(SO3Na)2)3]Cl2. Photochem Photobiol Sci 9(1):103–109

61. Borisov SM, Mayr T, Mistlberger G, Waich K, Koren K, Chojnacki P, Klimant I (2009) Precipitation as a simple and versatile method for preparation of optical nanochemosensors. Talanta 79(5):1322–1330. doi:10.1016/j.talanta.2009.05.041

62. Wu C, Bull B, Christensen K, McNeill J (2009) Ratiometric single-nanoparticle oxygen sensors for biological imaging. Angew Chemie Int Ed 48(15):2741–2745. doi:10.1002/anie.200805894

63. Wang X-d, Gorris HH, Stolwijk JA, Meier RJ, Groegel DBM, Wegener J, Wolfbeis OS (2011) Self-referenced RGB colour imaging of intracellular oxygen. Chem Sci 2(5):901–906

64. Zhdanov AV, Ogurtsov VI, Taylor CT, Papkovsky DB (2010) Monitoring of cell oxygenation and responses to metabolic stimulation by intracellular oxygen sensing technique. Integr Biol 2(9):443–451

65. Golub AS, Barker MC, Pittman RN (2007) PO2 profiles near arterioles and tissue oxygen consumption in rat mesentery. Am J Physiol Heart Circ Physiol 293(2):H1097–H1106. doi:10.1152/ajpheart.00077.2007

66. Golub AS, Pittman RN (2008) PO2 measurements in the microcirculation using phosphorescence quenching microscopy at high magnification. Am J Physiol Heart Circ Physiol 294(6):H2905–H2916. doi:10.1152/ajpheart.01347.2007

67. Golub AS, Tevald MA, Pittman RN (2011) Phosphorescence quenching microrespirometry of skeletal muscle in situ. Am J Physiol Heart Circ Physiol 300(1):H135–H143. doi:10.1152/ajpheart.00626.2010

68. Pittman RN, Golub AS, Carvalho H (2010) Measurement of oxygen in the microcirculation using phosphorescence quenching microscopy. Oxyg Transp Tissue XXXI. In: Takahashi E, Bruley DF (eds) Advances in experimental medicine and biology, vol 662. Springer, US, pp 157–162. doi:10.1007/978-1-4419-1241-1_22

69. Shonat RD, Kight AC (2003) Oxygen tension imaging in the mouse retina. Ann Biomed Eng 31(9):1084–1096. doi:10.1114/1.1603256

70. Lo L-W, Koch CJ, Wilson DF (1996) Calibration of oxygen-dependent quenching of the phosphorescence of Pd-meso-tetra (4-Carboxyphenyl) porphine: a phosphor with general application for measuring oxygen concentration in biological systems. Anal Biochem 236(1):153–160. doi:10.1006/abio.1996.0144

71. Estrada AD, Ponticorvo A, Ford TN, Dunn AK (2008) Microvascular oxygen quantification using two-photon microscopy. Opt Lett 33(10):1038–1040

72. Wilson DF, Vinogradov SA, Grosul P, Sund N, Vacarezza MN, Bennett J (2006) Imaging oxygen pressure in the rodent retina by phosphorescence lifetime. In: Cicco G, Bruley DF, Ferrari M (eds) Advances in experimental medicine and biology, vol 578. Springer, US, pp 119–124. doi:10.1007/0-387-29540-2_19

73. Diepart C, Verrax J, Calderon PB, Feron O, Jordan BF, Gallez B (2010) Comparison of methods for measuring oxygen consumption in tumor cells in vitro. Anal Biochem 396(2):250–256

74. Hynes J, Marroquin LD, Ogurtsov VI, Christiansen KN, Stevens GJ, Papkovsky DB, Will Y (2006) Investigation of drug-induced mitochondrial toxicity using fluorescence-based oxygen-sensitive probes. Toxicol Sci 92(1):186–200. doi:10.1093/toxsci/kfj208

75. Zhdanov AV, Favre C, O'Flaherty L, Adam J, O'Connor R, Pollard PJ, Papkovsky DB (2011) Comparative bioenergetic assessment of transformed cells using a cell energy budget platform. Integr Biol 3(11):1135–1142

76. Sakadzic S, Roussakis E, Yaseen MA, Mandeville ET, Srinivasan VJ, Arai K, Ruvinskaya S, Devor A, Lo EH, Vinogradov SA, Boas DA (2010) Two-photon high-resolution measurement of partial pressure of oxygen in cerebral vasculature and tissue. Nat Meth 7(9):755–759

77. Finikova OS, Lebedev AY, Aprelev A, Troxler T, Gao F, Garnacho C, Muro S, Hochstrasser RM, Vinogradov SA (2008) Oxygen microscopy by two-photon-excited phosphorescence. ChemPhysChem 9(12):1673–1679. doi:10.1002/cphc.200800296

78. Esipova TV, Karagodov A, Miller J, Wilson DF, Busch TM, Vinogradov SA (2011) Two new "protected" oxyphors for biological oximetry: properties and application in tumor imaging. Anal Chem. doi:10.1021/ac2022234

79. Napp J, Behnke T, Fischer L, Würth C, Wottawa M, Katschinski DM, Alves F, Resch-Genger U, Schäferling M (2011) Targeted luminescent near-infrared polymer-nanoprobes for in vivo imaging of tumor hypoxia. Anal Chem. doi:10.1021/ac201870b

80. Zhdanov AV, Ward MW, Taylor CT, Souslova EA, Chudakov DM, Prehn JH, Papkovsky DB (2010) Extracellular calcium depletion transiently elevates oxygen consumption in neurosecretory PC12 cells through activation of mitochondrial Na(+)/Ca(2+) exchange. Biochimica et biophysica acta 1797(9):1627-1637. doi:S0005-2728(10)00629-8[pii]10.1016/j.bbabio.2010.06.006

81. Zhdanov AV, Ward MW, Prehn JHM, Papkovsky DB (2008) Dynamics of intracellular oxygen in PC12 cells upon stimulation of neurotransmission. J Biol Chem 283(9):5650–5661. doi:10.1074/jbc.M706439200

82. O'Riordan TC, Zhdanov AV, Ponomarev GV, Papkovsky DB (2007) Analysis of intracellular oxygen and metabolic responses of mammalian cells by time-resolved fluorometry. Anal Chem 79(24):9414–9419. doi:10.1021/ac701770b

83. Zhdanov A, Dmitriev R, Papkovsky D (2011) Bafilomycin A1 activates respiration of neuronal cells via uncoupling associated with flickering depolarization of mitochondria. Cell Mol Life Sci 68(5):903–917. doi:10.1007/s00018-010-0502-8

84. Koo Y-EL, Cao Y, Kopelman R, Koo SM, Brasuel M, Philbert MA (2004) Real-time measurements of dissolved oxygen inside live cells by organically modified silicate fluorescent nanosensors. Anal Chem 76(9):2498–2505. doi:10.1021/ac035493f

85. Wotzlaw C, Bernardini A, Berchner-Pfannschmidt U, Papkovsky D, Acker H, Fandrey J (2011) Multifocal animated imaging of changes in cellular oxygen and calcium concentrations and membrane potential within the intact adult mouse carotid body ex vivo. Am J Physiol Cell Physiol. doi:10.1152/ajpcell.00508.2010

86. Dmitriev RI, Zhdanov AV, Ponomarev GV, Yashunski DV, Papkovsky DB (2010) Intracellular oxygen-sensitive phosphorescent probes based on cell-penetrating peptides. Anal Biochem 398(1):24–33

87. Fercher A, Borisov SM, Zhdanov AV, Klimant I, Papkovsky DB (2011) Intracellular O$_2$ sensing probe based on cell-penetrating phosphorescent nanoparticles. ACS Nano 5:5499–5508. doi:10.1021/nn200807g

88. Kondrashina AV, Dmitriev RI, Borisov SM, Klimant I, O'Brian I, Nolan YM, Zhdanov AV, Papkovsky DB (2012) A phosphorescent nanoparticle based probe for sensing and imaging of (intra)cellular oxygen in multiple detection modalities. Adv Funct Mater. doi:10.1002/adfm.201201387

89. Lakowicz JR (2006) Principles of fluorescence Spectroscopy, 3rd ed. Springer, 954 p

90. Periasamy A, Diaspro A (2003) Multiphoton microscopy. J Biomed Opt 8(3):327–328. doi:10.1117/1.1594726

91. Periasamy A, Elangovan M, Elliott E, Brautigan DL (2002) Fluorescence lifetime imaging (FLIM) of green fluorescent fusion proteins in living cells. Methods Mol Biol 183:89–100. doi:10.1385/1-59259-280-5:089

92. Xiang H, Zhou L, Feng Y, Cheng J, Wu D, Zhou X (2012) Tunable fluorescent/phosphorescent platinum(II) porphyrin–fluorene copolymers for ratiometric dual emissive oxygen sensing. Inorgan Chem. doi:10.1021/ic300040n

93. Tsai AG, Friesenecker B, Mazzoni MC, Kerger H, Buerk DG, Johnson PC, Intaglietta M (1998) Microvascular and tissue oxygen gradients in the rat mesentery. Proc Natl Acad Sci 95(12):6590–6595

94. Fercher A, O'Riordan TC, Zhdanov AV, Dmitriev RI, Papkovsky DB (2010) Imaging of cellular oxygen and analysis of metabolic responses of mammalian cells. Methods Mol Biol 591:257–273. doi:10.1007/978-1-60761-404-3_16

95. Becker W, Su B, Holub O, weisshart K (2010) FLIM and FCS detection in laser-scanning microscopes: Increased efficiency by GaAsP hybrid detectors. Microsc Res Tech 74(9):804–811. doi:10.1002/jemt.20959

96. Will Y, Hynes J, Ogurtsov VI, Papkovsky DB (2006) Analysis of mitochondrial function using phosphorescent oxygen-sensitive probes. Nat Protoc 1(6):2563–2572. doi:nprot.2006.351[pii]10.1038/nprot.2006.351

97. Zitova A, Hynes J, Kollar J, Borisov SM, Klimant I, Papkovsky DB (2010) Analysis of activity and inhibition of oxygen-dependent enzymes by optical respirometry on the LightCycler system. Anal Biochem 397(2):144–151

98. Papkovsky DB, O'Riordan T, Soini A (2000) Phosphorescent porphyrin probes in biosensors and sensitive bioassays. Biochem Soc Trans 28(2):74–77

99. O'Mahony FC, Papkovsky DB (2006) Rapid high-throughput assessment of aerobic bacteria in complex samples by fluorescence-based oxygen respirometry. Appl Environ Microbiol 72(2):1279–1287. doi:10.1128/aem.72.2.1279-1287.2006

100. O'Flaherty L, Adam J, Heather LC, Zhdanov AV, Chung YL, Miranda MX, Croft J, Olpin S, Clarke K, Pugh CW, Griffiths J, Papkovsky D, Ashrafian H, Ratcliffe PJ, Pollard PJ (2010) Dysregulation of hypoxia pathways in fumarate hydratase-deficient cells is independent of defective mitochondrial metabolism. Hum Mol Genet 19(19):3844–3851. doi:ddq305 [pii] 10.1093/hmg/ddq305

101. O'Hagan KA, Cocchiglia S, Zhdanov AV, Tambuwala MM, Cummins EP, Monfared M, Agbor TA, Garvey JF, Papkovsky DB, Taylor CT, Allan BB (2009) PGC-1alpha is coupled to HIF-1alpha-dependent gene expression by increasing mitochondrial oxygen consumption in skeletal muscle cells. Proc Natl Acad Sci U S A 106(7):2188–2193. doi:0808801106 [pii] 10.1073/pnas.0808801106

102. Frezza C, Zheng L, Tennant DA, Papkovsky DB, Hedley BA, Kalna G, Watson DG, Gottlieb E (2011) Metabolic profiling of hypoxic cells revealed a catabolic signature required for cell survival. PLoS ONE 6(9):e24411. doi:10.1371/journal.pone.0024411

103. Semenza GL (2010) Oxygen homeostasis. Wiley Interdiscip Rev Sys BiolMed 2(3):336–361. doi:10.1002/wsbm.69

104. Wong C, Zhang H, Gilkes D, Chen J, Wei H, Chaturvedi P, Hubbi M, Semenza G (2012) Inhibitors of hypoxia-inducible factor 1 block breast cancer metastatic niche formation and lung metastasis. J Mol Med 90:803–815. doi:10.1007/s00109-011-0855-y

105. Ferrick DA, Neilson A, Beeson C (2008) Advances in measuring cellular bioenergetics using extracellular flux. Drug Discov Today 13(5–6):268–274. doi:10.1016/j.drudis.2007.12.008

106. Huppert TJ, Allen MS, Benav H, Jones PB, Boas DA (2007) A multicompartment vascular model for inferring baseline and functional changes in cerebral oxygen metabolism and arterial dilation. J Cereb Blood Flow Metab Off J Int Soc Cereb Blood Flow Metab 27(6):1262–1279

107. Fang Q, Sakadzic S, Ruvinskaya L, Devor A, Dale AM, Boas DA (2008) Oxygen advection and diffusion in a three- dimensional vascular anatomical network. Opt Express 16(22):17530–17541

108. Zheng L, Golub AS, Pittman RN (1996) Determination of PO2 and its heterogeneity in single capillaries. Am J Physiol Heart Circ Physiol 271(1):H365–H372

109. Yaseen MA, Srinivasan VJ, Sakadzic S, Wu W, Ruvinskaya S, Vinogradov SA, Boas DA (2009) Optical monitoring of oxygen tension in cortical microvessels with confocal microscopy. Opt Express 17(25):22341–22350. doi:10.1364/OE.17.022341190652 [pii]

110. Howlett RA, Kindig CA, Hogan MC (2007) Intracellular PO2 kinetics at different contraction frequencies in Xenopus single skeletal muscle fibers. J Appl Physiol 102(4):1456–1461. doi:10.1152/japplphysiol.00422.2006

111. Zhang S, Hosaka M, Yoshihara T, Negishi K, Iida Y, Tobita S, Takeuchi T (2010) Phosphorescent lightâ€"emitting iridium complexes serve as a hypoxia-sensing probe for tumor imaging in living animals. Can Res 70(11):4490–4498. doi:10.1158/0008-5472.can-09-3948

112. Robiolio M, Rumsey WL, Wilson DF (1989) Oxygen diffusion and mitochondrial respiration in neuroblastoma cells. Am J Physiol Cell Physiol 256(6):C1207–C1213

113. Mik EG, Ince C, Eerbeek O, Heinen A, Stap J, Hooibrink B, Schumacher CA, Balestra GM, Johannes T, Beek JF, Nieuwenhuis AF, van Horssen P, Spaan JA, Zuurbier CJ (2009) Mitochondrial oxygen tension within the heart. J Mol Cell Cardiol 46(6):943–951

114. van Dongen JT, Gupta KJ, Ramírez-Aguilar SJ, Araújo WL, Nunes-Nesi A, Fernie AR (2011) Regulation of respiration in plants: A role for alternative metabolic pathways. J Plant Physiol 168(12):1434–1443

115. Zitova A, O'Mahony FC, Cross M, Davenport J, Papkovsky DB (2009) Toxicological profiling of chemical and environmental samples using panels of test organisms and optical oxygen respirometry. Environ Toxicol 24(2):116–127. doi:10.1002/tox.20387

116. Lo JF, Wang Y, Blake A, Yu G, Harvat TA, Jeon H, Oberholzer J, Eddington DT (2012) Islet preconditioning via multimodal microfluidic modulation of intermittent hypoxia. Anal Chem 84(4):1987–1993. doi:10.1021/ac2030909

Chapter 2
O_2 Analysis on a Fluorescence Spectrometer or Plate Reader

Alexander V. Zhdanov, James Hynes, Ruslan I. Dmitriev
and Dmitri B. Papkovsky

Abstract In this chapter, the use of Pt-porphyrin-based extracellular and intracellular O_2 sensing probes (ecO$_2$ and icO$_2$) on a TR-F plate reader format is described and critically assessed. The principles underpinning extracellular measurement are outlined and the assessment of biological oxygen consumption in prokaryotic cells, eukaryotic cells and small organisms is outlined along with a description of how such measurements can contribute to the development of a detailed picture of cell metabolism. The use of a mathematical model describing the distribution of local O_2 gradients within biological samples is also described. Finally, the principles of icO$_2$ sensing are discussed and some short case studies are provided for demonstrating the utility of icO$_2$ probes for the monitoring of cellular function and metabolic perturbations and how such measurements can be allied to other bioenergetic markers to generate a more complete picture of metabolic status.

Keywords Oxygen · Respiration · Cell-based assays · Phosphorescent Probes · Pt-porphyrins · Time-Resolved Fluorescence · Oxygen-Sensitive Probes · Screening

2.1 Introduction

As outlined in Chap. 1, the simplest approach to the assessment of biological O_2 consumption by luminescence quenching is to introduce an O_2-sensitive material (a dispensable probe [1] or a solid-state sensor [2]) into the medium surrounding of the test specimen, measure its luminescent signal on a conventional fluorescence spectrometer or plate reader and then relate this signal to O_2 concentration or consumption rate (OCR). Extracellular O_2 (ecO$_2$) probes allow simple, parallel OCR

D. B. Papkovsky et al., *Phosphorescent Oxygen-Sensitive Probes*,
SpringerBriefs in Biochemistry and Molecular Biology,
DOI: 10.1007/978-3-0348-0525-4_2, © The Author(s) 2012

measurement in multiple samples facilitating the assessment of metabolic impact of biological processes such as cell transformation genetic manipulation, drug treatment or culture conditions. Measurements are typically performed on adherent or suspension cell populations in microplate format. The method can also be applied to other specimens including aquatic organisms and model animals (such as *C. elegans*, *Artemia*, *Zebrafish*, *Daphnia*) [3, 4], and on more specialised measurement formats such as customised microplates [5], capillary cuvettes [6] or fluidic biochips [7].

For specific research questions, information on the O_2 concentration within the cell rather in the medium surrounding the cell can be particularly informative and provide a different insight into cellular function [8]. This can be achieved using *intracellular* O_2 (icO_2) probes representing (macro)molecular [9] or nano-particulate [10] structures delivered into the cell either by means of commercial transfection reagents [11], microinjection or by self-loading via endocytosis pathways [10, 12–14]. The latter approach relies on cell-penetrating O_2 probes supersedes the other strategies due to its convenience, reproducibility and efficiency. Optical measurements with loaded cells can be conducted over several hours [15], allowing for real-time monitoring of cell oxygenation levels at different conditions and levels of atmospheric O_2, and of rapid transient respiratory responses to cell treatment [16, 17].

Both techniques are simple, contactless, quantitative and facilitate high-throughput plate reader-based analysis of OCR or icO_2, providing useful and physiologically relevant information on cellular function. The optical detector interrogates the probe from outside the sample making these platforms flexible and non-invasive. Best performance is achieved when using high-sensitivity multi-label time-resolved fluorescence (TR-F) readers in lifetime measurement mode such as RLD (see Chap. 1). Such instruments are produced by a number of vendors (e.g., FLUOstar and POLARStar Omega® instruments from BMG Labtech GmbH, Germany and the Victor® family from PerkinElmer, USA), they are available in many life sciences labs and can be used without major modifications with several existing Pt-porphyin-based O_2 sensing probes (also produced commercially). These instruments also allow for multiplexed and parallel use of O_2 probes with other compatible probes and assays [18].

While the imaging-based measurements (outlined in detail in Chap. 3) offer considerable information on intracellular oxygen facilitating single cell detalisation and high-resolution 2D and 3D O_2 mapping, such systems suffer from some drawbacks with regard to sample throughput and a requirement for more sophisticated, often custom-modified equipment. In contrast, plate-reader-based O_2 analysis can facilitate intracellular analysis with much higher sample throughput using less sophisticated broadly available instrumentation, albeit at the expense of a degree of detail.

In this chapter, we describe the various uses of Pt-porphyrin-based extracellular and intracellular O_2 probes, in conjunction with plate-reader-based analysis across various biological systems. A significant number of applications developed in our lab and experimental studies performed in collaboration with academic and industrial partners are presented. Critical technical considerations are outlined, the potential for parallel measurements with other added-value techniques is explained in detail, with the aim of providing the reader with a more comprehensive under-standing of the capabilities and limitations of these technologies and a guide as to how to apply these techniques in their own research.

2.2 Bioassays Performed with ecO$_2$ Probes

Extracellular probes are typically deployed in the test medium and report on the rate at which biological O$_2$ consumption depletes the O$_2$ concentration within the sample. When devising an appropriate set-up for such measurements, the two main considerations are the OCR within the sample and the rate at which ambient O$_2$ diffuses back into the system. The most convenient format for such analysis is the standard microtiter plate. This approach can be applied to any test system which produces a detectable level of O$_2$ consumption with examples including isolated mitochondria [19], enzymes [6], yeast and prokaryotic suspensions [20], adherent eukaryotic cells [21], spheroids and small aquatic organisms [4]. The dissolved O$_2$ gradient generated while O$_2$ is consumed within the well drives the diffusion of ambient O$_2$ into the sample at both the liquid–air interface and through the body of the microplate (common plastic materials such as polystyrene are rather permeable to O$_2$). Measurement, therefore, usually necessitates the use of a seal, often a mineral oil overlay, to slow such back diffusion. After an equilibration phase, either a steady state is reached or all available O$_2$ are depleted.

The flexibility associated with ecO$_2$ probes also facilitates measurements in instances where the amount of analysed biomaterial is limited. This is achieved through the use of narrow bore glass capillaries where high sensitivity measure-ment can be conducted in volumes as low as several microliters [6], and further volume reduction is possible using specially designed devices made of O$_2$-impermeable materials [5, 7]. The capillary cuvette provides considerable sensi-tivity due to the limited surface area available and thick barrier of liquid for O$_2$ back diffusion and can be applied to the analysis of multiple samples on the commercial LightCycler® system [6] which was originally developed for quanti-tative PCR. Ground capped quartz or glass cuvettes provide hermetic seal which prevents the ingress of ambient O$_2$ and allows accurate and sensitive O$_2$ analysis on a TR-F reader or even intensity-based spectrometer, as outlined in Chap. 1 (Fig. 1.5).

2.2.1 Analysis of Isolated Mitochondria

One of the areas, where the use of phosphorescence based ecO$_2$ probes have seen widespread adoption, is in the assessment of mitochondria activity, specifically as a high-throughput method of analysing the metabolic implications of drug treat-ment. OCR measurements are favoured in this regard as they assess directly the activity of the electron transport chain (ETC) and are, therefore, highly sensitive to perturbations in mitochondrial function. Much of this work is conducted on the organelles isolated from rat liver or heart and, as with any organelle-based assays, a key parameter is the quality of the mitochondrial preparation [19, 22]. These measurements are particularly useful where mechanistic information is sought on the mode of action of the mitochondria perturbation as it assures free access of the compound to the machinery of the mitochondria, removing the possible complications with transportation across cell membrane. Detailed mechanistic investigations can be achieved through the use of substrates which feed reducing equivalents to specific complexes along the ETC.: glutamate/malate, succinate and ascorbate/TMPD (N,N',N',N'-tetramethyl-p-phenylenediamine) for complexes I, II and III, respectively. Fatty acid oxidation pathway can also be probed through the use of palmitate as a substrate.

Traditionally, such analyses were achieved using standard polarography, however, limited throughput has restricted this adoption. The necessary throughput is achieved using a water-soluble O$_2$ probe on a standard microplate format [19], which allows for the generation of dose response data and analysis of multiple substrates in both the presence and absence of ADP. Sample data are presented in Fig. 2.1 illustrating phosphorescence intensity profiles which reflect O$_2$ profiles and OCRs produced by the samples of rat liver mitochondria. Measurement in the presence of ADP allows the detection of inhibitors of ETC function as seen by the antimycin A (Ant A) treatment in Fig. 2.1a, while rotenone, a potent complex I inhibitor does not cause inhibition as succinate is the substrate feeding reducing equivalents via complex II and bypassing the site of inhibition. Rotenone does, however, show inhibition where glutamate/malate are substrates (Fig. 2.1b) as in this instance, reducing equivalents are fed through complex I. Measurements in Fig. 2.1b are conducted in the absence of ADP allowing the detection of uncoupling as shown by FCCP treatment. Such measurements can also be used to assess the involvement of the mitochondrion in various disease states [23, 24].

2.2.2 Analysis of Mammalian Cell Respiration

While analysis in the isolated organelle is particularly advantageous in specific circumstances, for many applications it is preferable to assess mitochondrial function within the whole cell, where it is fully integrated with the various secondary metabolic controls and regulatory mechanisms while also ensuring the integrity of the cellular ultrastructure and related signalling. There has been a growing realisation over recent years of the importance of this integration, to the extent that mitochondria are no longer considered as unifunctional free standing organelles but as a complex dynamic reticulum [25]. It is clear, therefore, that cell-based assessment of mitochondrial function is of particular importance, especially for the studies of the compounds or treatments that obstruct oxidative metabolism. However, many standard assays measure end-points which are secondary to the

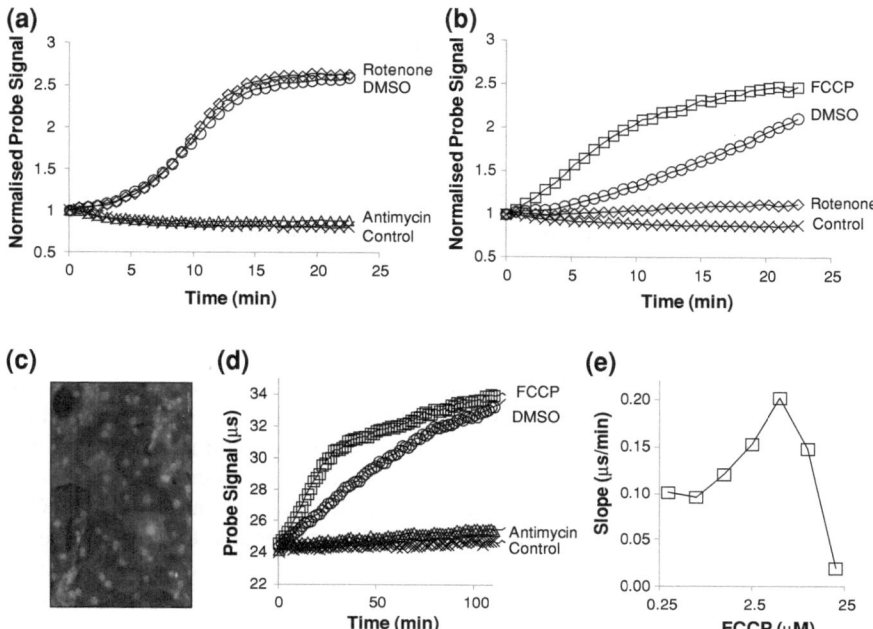

Fig. 2.1 Respiration profiles for isolated mitochondria (**a, b**) and whole cells (**d, e**) measured with the ecO$_2$ probe MitoXpress®. **a** Normalised intensity profiles for isolated mitochondria on Succinate in the presence of ADP showing ETC inhibition. **b** Isolated mitochondria on Glutamate/Malate in the absence of ADP showing ETC inhibition and uncoupling. **c** Immuno-fluorescence of Cor.At mESC derived cardiomyocytes demonstrating s-actin (*green*) contractile filaments and connexin 43 (*orange*) gap junctions between cardiomyocytes. **d** Lifetime-based respiration profiles of Cor.At cell showing both ETC uncoupling and inhibition (Ant A). **e** FCCP dose response whereby an initial increase in OCR is followed by a significant reduction in ETC activity caused by non-specific cell damage. Image used with permission from Axiogenesis AG (Germany)

activity of the mitochondria. Under standard culture conditions, many cell lines exhibit the capacity to maintain cellular ATP supply via increased glycolytic flux despite the complete shutdown of oxidative phosphorylation (OxPhos) metabolism, a phenomenon termed the Crabtree effect [26]. In such instances, if secondary parameters, such as ATP, or the reductive capacity of the cell as measured by MTT or Alamar blue assays, are taken as an indication of mitochondrial health, this can lead to a considerable underestimation of the level of metabolic inhibition [26]. For this reason, direct OCR measurement is particularly valuable as it facilitates a direct analysis of the mitochondrial function and is not susceptible to such misinterpretation. OCR measurement in whole cells is conducted in a manner similar to that used for isolated mitochondria where the rate of O$_2$ depletion within the sample is monitored over time [21].

The ability to interrogate mitochondrial function while maintaining cellular ultrastructure and related signalling is evident in the assessment of stem cell-derived cardiomyocytes. Such cell cultures are often favoured over primary cells as they circumvent the isolation and purity difficulties often associated with primary cells. Figure 2.1c–e illustrates the measurement of relative OCR for Cor.At cardiomyocytes (Axiogenesis AG, Germany), which are produced from mouse embryonic stem (ES) cells through in vitro differentiation and, similar to primary cells, have limited proliferative capacity. These cells exhibit a high energy demand, express functional cardiomyocyte ion channels and receptors, and a stable beating signal. They are puromycin resistant thereby allowing the generation of a homogenous and reproducible cell system [27]. Their ultrastructural complexity is illustrated in Fig 2.1c, showing cells after a 4 week culture period with s-actin stained green, highlighting the contractile filaments and Connexin 43 stained orange demonstrating gap junctions between cardiomyocytes. The metabolic impact of compounds known to inhibit electron transport chain function is presented in Fig. 2.1d, where both Ant A and the FoF1 ATP synthase inhibitor oligomycin induce an immediate inhibition of OCR. In contrast, treatment with FCCP causes an uncoupling related increase in OCR, the associated dose–response relationship is presented in Fig. 2.1e. Cellular OCR increases until a critical FCCP concentration is reached after which point, OCR declines due to cell damage.

This type of analysis with ecO$_2$ probes has been applied to many cell types including primary neurons [28], primary rat hepatocytes [21], myoblasts [29], cancer cells [30] and examining phenomena such as pharmacological perturbation of the ETC. activity [31], genetic modulation of ETC complexes [32] and the role of PGC-1α and PNC1 in mitochondrial biogenesis and function [29, 30]. This approach has also been perused as a diagnostic tool for the evaluation of suspected mitochondrial disorders using digitonin permeabilised patient-derived fibroblasts. Permeabilisation allows direct access of compound or substrate to the mitochondrial network facilitating an analysis of the entire OxPhos apparatus in whole cells and allowing comparison with data from human-derived isolated mitochondria. Such a comparison aids the differentiation of primary and secondary mitochondriopathies. This approach is also applicable to the screening of cybrid clones for mitochondrial dysfunction related to mtDNA mutation [33].

Despite advances in the development of biological models such as those described above, the standard 2D cultured cells still lack the biological complexity necessary to investigate certain biochemical processes. This is particularly relevant in tumour biology where the tumour microenvironment often contains hypoxic regions which are thought to be relevant to tumour aggressiveness and response to particular therapies. Multi-cellular spheroids offer an alternative to 2D cultures for the investigation of such biology where a spherical symmetry and more complex cell-to-cell interactions result in a more relevant biological model. Metabolism and the OCR is an important parameter, particularly considering the limited diffusion of O$_2$ and nutrients across the spheroid.

Such measurements can also be performed in microplate format whereby multiple spheroids are placed in a test well and measured as outlined above for differing cell lines, culture conditions and substrates. Figure 2.2 shows relative OCRs from spheroids of the human glioblastoma cell line U87MG and the human

Fig. 2.2 Relative OCRs for U87MG, MO59 K spheroids pre-conditioned to glucose/pyruvate media for 4–6 h. Measurements conducted with spheroids from 1 mm average diameter at 3 spheroids per well of 96 well plate. (Data with permission from Michelle Potter, and Dr. Karl Morten, University of Oxford, U.K)

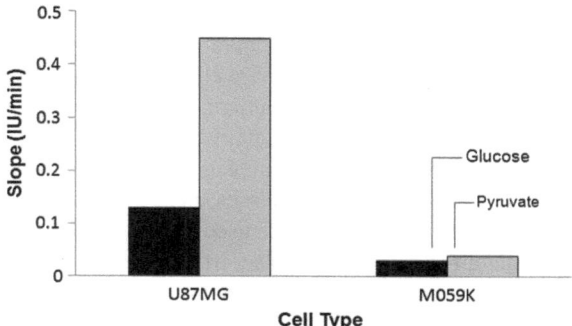

glioma cell line MO59 K, and the effect of glucose and pyruvate. U87MG spheroids are seen to be considerably more aerobic, and they show a significant increase in OCR when preconditioned in pyruvate compared to conventional glucose culture. A difficulty with the measurement of spheroids is the inherent variability that can exist in the amount of biomaterial in each replicate well due to variation in spheroid size. This can be addressed through the use of dedicated spheroid culture plates where the spheroids developed in a 'hanging drop' [34]. Using dedicated low volume plates, OCR of two or even one spheroid can be measured.

2.2.3 Analysis of Drug-Induced Mitochondrial Dysfunction by Parallel Measurement of OCR and Glycolytic Flux

While knowledge of cellular OCR provides direct information on OxPhos, an accompanying analysis of glycolysis, the other main ATP generating pathway in cells can also be valuable. This can be achieved by measuring the rate of extra-cellular acidification (ECA) of the media in which the cells grow. In an open system, the vast majority of this acidification is due to the production of lactic acid. The ECA can be followed using fluorescent pH-sensitive probes which, when spectrally compatible, facilitate true multiplexing where both probes are deployed together thereby reporting on both oxidative and glycolytic metabolism in the same test well. For example, the pH-sensitive probe pH-Xtra, which comprises a long decay narrow band emitting Eu(III)-chelate, can be multiplexed with Pt-porphyrin-based MitoXpress® ecO₂ probe (produced by Luxcel Biosciences, Ireland), to achieve simultaneous readout of O_2 and H^+ concentrations within the same sample with practically no cross-sensitivity [18]. pH-Xtra probe also changes its emission lifetime as a function of pH, so that both probes can be measured in RLD mode on standard TR-F readers.

It should be noted that, while such multiplexing is possible, from a data interpretation perspective, it is often simpler to conduct these measurements in parallel as the seal required for OCR measurements results in the trapping of Krebs cycle derived CO_2 resulting in additional media acidification unrelated to glycolytic flux. A parallel measurement whereby OCR is measured in sealed wells and ECA is measured in unsealed wells is therefore preferable.

The compensatory mechanism outlined above, whereby cells can circumvent mitochondrial insult to maintain cellular ATP by increasing glycolytic flux, is particularly relevant to the investigation of drug-induced mitochondrial dysfunction. It can, however, be harnessed for screening whereby OCR is used as the primary indicator and ECA is used as a confirmatory parameter in both 96 and 384 well plate format. In such case, true mitochondria inhibitors will cause a decrease in O_2 consumption and an accompanying increase in ECA while general cell toxicity will cause a decrease in both parameters [18] as no compensation will be possible. Uncouplers would be expected to increase in both ECA and OCR. At the same time, it is important to be aware of the bell shaped dose response of ten associated with such uncouplers when interpreting such data (Fig. 2.1e). Figure 2.3 illustrates how this approach can be applied. A small library of drugs is screened at fixed concentration with data presented as to the strength of the response illicited. As expected, known inhibitors populate the top left quadrant indicating ETC inhibition and increased glycolytic flux while non-mitochondrial toxicants and those that inhibit upstream of glycolysis populate the bottom left quadrant. Uncouplers are expected to populate the top right quadrant indicating increased ETC and glycolytic flux. Compounds with no detectable effect at the test concentration populate the region around the origin. This approach is currently being expanded to include full dose response analysis and to examine the relationship between such metabolic screening and other available measurements of mitochondrial dysfunction.

2.2.4 Multi-Parametric Assessment of 'Cell Energy Budget'

The energy flux compensation illustrated in Sect. 2.2.3 has recently been developed further into the so-called 'Cell Energy Budget' (CEB) concept. CEB is particularly useful for the analysis of perturbed metabolism, mitochondrial and glycolytic disorders, which upon energy stress may lead to cellular malfunction and progression of various diseases [35–37].

OxPhos and glycolysis linked by the Krebs cycle serve as the main sources of ATP in eukaryotes, these pathways are mutually regulated in order to maintain optimal energy balance in the cell. The relative contribution of these pathways to cell bioenergetics varies broadly for different cell types and conditions. Other metabolic pathways also contribute to cell bioenergetics [38, 39], including β-oxidation of fatty acids, the pentose phosphate pathway and glutaminolysis, the latter is particularly important for cancer cell metabolism [40]. Normally, each of the main pathways has a considerable spare capacity outside the normal physiological range, thus allowing the cells to maintain stable ATP levels in stress conditions. However, metabolic abnormalities and mutations in genes linked directly or indirectly to OxPhos,

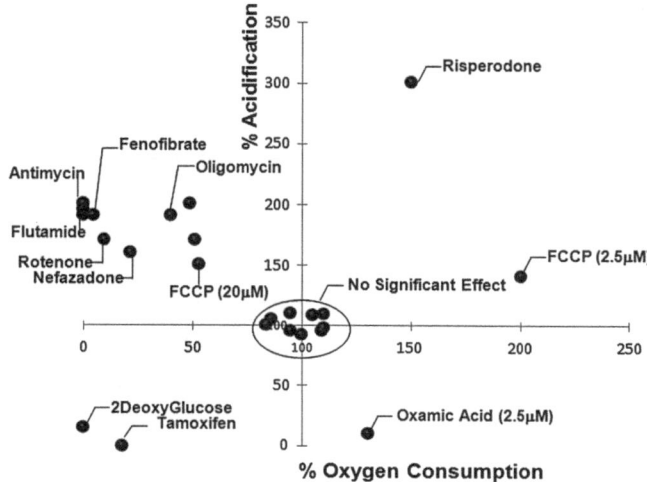

Fig. 2.3 Single concentration data analysis matrix assessing the effect of a panel of drugs on HepG2 cell metabolism including compounds of known mode of action

glycolysis, glutaminolysis or the Krebs cycle may alter their 'normal' contribution to ATP fluxes and reduce spare capacity [30, 40, 41]. For example in conditions of sustained excitation, shortage of nutrients or imbalance in energy-generating pathways the cells may be plunged into an energy crisis or become vulnerable to stress factors.

The main energy-generating pathways in eukaryotic cells and strategy for multi-parametric CEB assessment are shown schematically in Fig. 2.4. Their spare capacity can be probed pharmacologically or by limiting the corresponding substrates necessary for ATP production. Thus, in normal cells, total ATP levels remain unchanged for many hours when OxPhos is repressed by mitochondrial uncoupling [15] or when glycolysis is inhibited by substitution of glucose with galactose [26]. In such manner, spare capacity of different pathways and ability of the cells to utilise alternative ATP-producing pathway(s) can be quantified with appropriate bioassays.

In context of general cell metabolism, ATP fluxes through OxPhos and glycolysis, as well as the activity of the Krebs cycle, are more informative as direct bioenergetic parameters than cellular redox state, Ca^{2+} and NAD(P)H levels, mitochondrial membrane potential ($\Delta\Psi$m) and pH. Various technological platforms have been introduced recently for comparative quantitative analysis of the key metabolic parameters of the cells. The automated Extracellular Flux Analyser can accurately measure cellular OCR and ECA in a special microplate with built-in solid-state O$_2$ and pH sensors [43]; however, it is not able to differentiate between glycolytic and non-glycolytic contribution to the ECA. The more flexible platform based on the long decay emitting O$_2$ and pH sensitive probes overcomes this limitation [18, 44]. Using standard microtiter plates and TR-F plate reader detection, it can measure lactate in unsealed (L-ECA, CO$_2$ escapes from the

Fig. 2.4 Representation of
the main energy-generating
pathways in aerobic cells and
strategy for CEB assessment
showing the interrelationship
of different pathways (*grey
gradient color* and *white
arrows*), ranges of
physiological activity (*solid
boxes*) and spare capacity
(*dark arrows*). The bioassays
used in quantitative CEB
analysis are shown in grey
boxes. Adapted from [42]

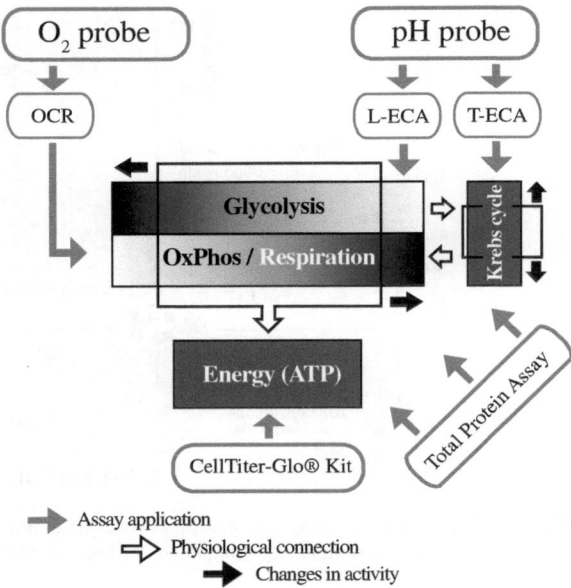

medium) and lactate plus CO_2 in sealed samples (total, T-ECA), thus discriminating between the glycolytic and non-glycolytic ECA components. Together with the measurement of OCR and total ATP levels, this CEB platform provided a detailed assessment of the contribution of each of the three main energy-generating pathways [42]. Notably, the main producer of CO_2 in the cell is the Krebs cycle, while the pentose phosphate pathway and pyruvate/malate cycle also produce CO_2 [45, 46].

To assess CEB with a minimal set of experiments, measurement of the cellular OCR, T-ECA, L-ECA and ATP values is conducted under both resting conditions and upon double treatment with FCCP/oligomycin. Inhibition of OxPhos by oligomycin can be used to probe mitochondrial respiration uncoupled from ATP production; however, oligomycin also increases proton motive force across the mitochondrial inner membrane and elevates the rate of proton leak, thus making the results difficult to interpret. In contrast, oligomycin/FCCP treatment strongly activates the OCR due to uncoupling and prevents the production and hydrolysis of ATP by Complex V. To compensate for decreased flux of mitochondrial ATP, the cells are forced to increase their glycolytic ATP production.

Individual metabolic assays and organisation of the CEB assessment are described in Table 2.1. The OCR and T-ECA are measured in sealed samples covered with oil to prevent back diffusion of ambient O_2 and escape of CO_2. They both can be measured in one kinetic assay on the first plate, and values expressed as relative changes of primary luminescent parameters over time, i.e. intensity or lifetime slopes. Similarly, the L-ECA and total ATP are both measured in unsealed samples exposed to ambient air. These assays can be performed sequentially on the second plate: after the completion of the kinetic L-ECA assay, total ATP is measured by cell lysis and end-point measurement of bioluminescent signals.

To reduce experimental error associated with different batches of cells and plates, resting and FCCP/oligomycin uncoupled cells are recommended to measure on the same plate. For accurate cross-comparison of different cell types, measured raw values of bioenergetic parameters should be normalised for different biomass content in corresponding samples, which can be determined in a separate experiment by measuring total protein concentration. In this manner, up to 96 or even 384 samples can be analysed in parallel on the same plate, producing comprehensive sets of data for different conditions, treatments and cell types. Necessary repeats, blanks and positive controls (reference or untreated cells) are also incorporated, as required.

The CEB approach was successfully applied to analyse the role of fumarate hydratase enzyme (FH) in cell energy production. Inactivating mutations of the gene encoding FH are known to cause a number of human cancers including hereditary leiomyomatosis and renal cell cancer (HLRCC) [47]. FH is localised predominantly in the mitochondria and at lower levels in the cytosol [48], and catalyses conversion of fumarate to malate [49]. This reaction is a critically important step of the Krebs cycle and determines the respiratory activity of the cell. FH deficiency in mouse embryonic fibroblasts (MEFs) strongly affects mitochondrial function leading to reduced mitochondrial respiration and elevated glycolysis [41].

This was further investigated by a comparative CEB assessment of the wild type (WT, Fh1$^{+/+}$) and Fh1$^{-/-}$ knockout (KO) MEFs. The KO cells were significantly smaller, so they were seeded at higher numbers than WT cells and for proper comparison the results were normalised for total protein content. OCR measurements with resting cells revealed a dramatic difference in OCR between the WT and KO cells (Fig. 2.5a). Upon uncoupling the OCR in WT cells increased by approximately twofold, while in KO cells it did not change indicating their low spare respiratory capacity [42]. ATP levels in KO were also lower than in WT cells (Fig. 2.5b), but upon uncoupling they stayed unchanged in both cell lines. As shown by L-ECA data, glycolysis of resting KO cells was increased $\sim$2-fold compared to the WT (Fig. 2.5c): 1.25 pH unit/min/10^6 cells. Upon uncoupling, L-ECA in WT cells increased $\sim$2-fold (shown by the arrow) to compensate for the loss of OxPhos flux, whereas in KO cells L-ECA remained unchanged. This is because most of the ATP in KO cells was already produced through glycolysis. The difference between L-ECA and T-ECA in KO cells was small due to a smaller contribution of the Krebs cycle in CO$_2$ production. The large decrease in OCR/L-ECA ratio for KO cells reflects the shift in CEB toward glycolytic energy production (Fig. 2.5d). This study demonstrates practical use of quantitative CEB assessment and O$_2$ monitoring as one of its key parameters.

Table 2.1 Proposed layout and description of the panel of metabolic assays used for CEB assessment. Multiplexable assays are colour coded and assigned plate number

Assay No	Description	Format	Samples	Measured parameters	Plate No.
1.	OxPhos activity Kinetic assay with MitoXpress® Probe (340/650 nm)	Sealed samples in 96/384 WP	Resting and uncoupled cells Panels of cells, different conditions and treatments	Respiration/OCR: R$_b$ and R$_{max}$	1
2.	Glycolytic Flux Kinetic assay with pH-Xtra Probe (340/615 nm)	Unsealed samples in 96/384 WP	Resting and uncoupled cells Panels of cells, different conditions and treatments	L-ECA G$_b$ and G$_{max}$	2
3.	Krebs Cycle Activity Kinetic assay with pH-Xtra Probe (340/615 nm)	Sealed samples in 96/384 WP	Resting and uncoupled cells Panels of cells, different conditions and treatments	T-ECA vs L-ECA K$_b$ and K$_{max}$	1
4.	ATP Assay, CellTiter-Glo® Kit (Promega)—BL measurement after cell lysis (end point, after assay 2)	Unsealed samples in 96/384 WP	Resting and uncoupled cells Panels of cells, different conditions and treatments	Total cellular ATP levels	2
5.	BCA™ Protein Assay—total protein measurement after cell lysis with convenient buffer containing detergent (end-point)	Unsealed samples in 96/384 WP	Different cells, long-term culture or treatments	Relative biomass content	3

Note R$_b$, R$_{max}$, G$_b$, G$_{max}$, K$_b$ and K$_{max}$—basal and maximal capacity of respiration (R), glycolysis (G) and Krebs cycle (K), respectively. Spare capacity is the difference of the two values

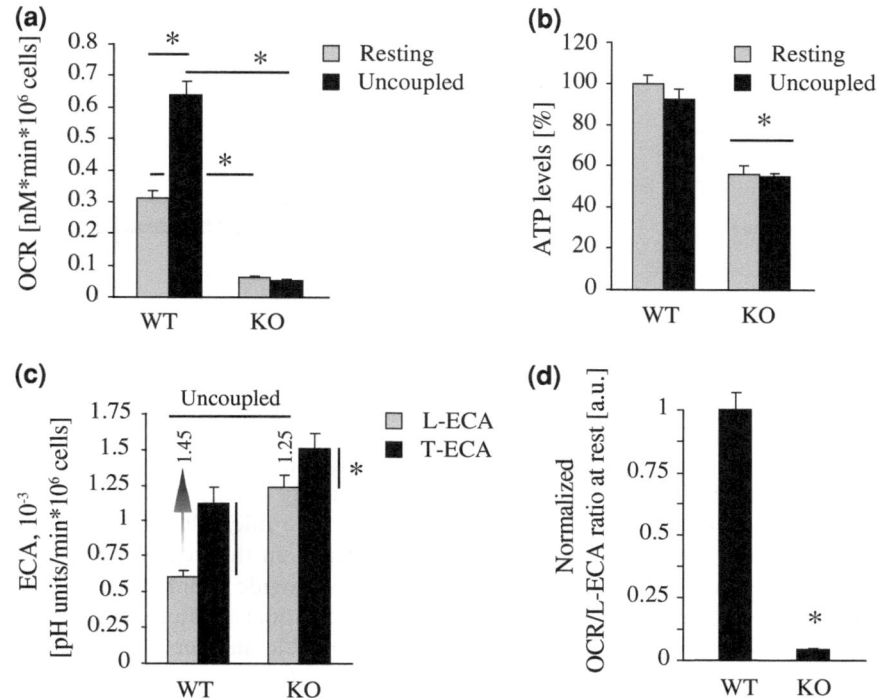

Fig. 2.5 Glycolytic shift in CEB associated with FH deficiency in MEF cells. **a** OCRs for the WT and KO MEF cells at rest and upon uncoupling with 1 µM FCCP/10 µM oligomycin. **b** Total ATP levels. **c** L-ECA and T-ECA for the resting cells. Arrows show their increase upon uncoupling. **d** OCR/L-ECA ration is dramatically reduced in KO cells. Asterisks indicate significant differences. Transformed FH cells were kindly provided by Julie Adam and Patrick Pollard, University of Oxford, UK

2.2.5 Monitoring of Oxygenation and Cell Respiration in Microfluidic Biochips

A number of microfluidic chips and microchamber devices developed for bio-logical applications including cell-based assays are now produced commercially. Such devices are often characterised by restricted permeation of atmospheric O$_2$ through biochip material, and limited diffusion of dissolved O$_2$ through their long and narrow-bore flow channels. When respiring cells and other biological samples are placed in such devices, control of oxygenation (and other relevant parameters such as pH, CO$_2$) becomes very important. Unfortunately, such control systems are not always considered and implemented by either the developers or end-users of such biochips.

Among the critical factors that determine the behaviour of the cells in biochip systems are cell density, O$_2$ availability, cell stress during the seeding and

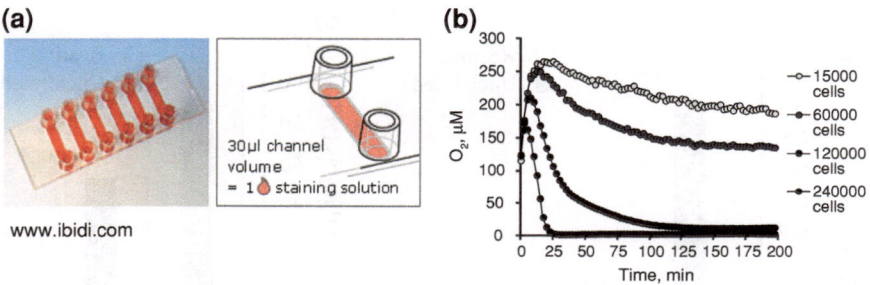

Fig. 2.6 Monitoring of cell deoxygenation with μ-slide VI$^{0.4}$ microchambers (Ibidi). **a** - images of the device. **b** - time profiles of oxygenation of adherent non-differentiated PC12 cells seeded at indicated numbers (per chamber, in 30 μL volume) and measured with the help of extracellular probe MitoXpress$^{®}$ on Victor2 reader

attachment (stopped flow conditions), changes in cell respiration rate and pro-longed measurements under static or limited O$_2$ supply conditions. These factors may cause rapid depletion of dissolved O$_2$ within the microchamber leading to hypoxic/anoxic shock, altered function or cell death. In this regard, O$_2$ sensing probes and phosphorescence quenching techniques provide simple tools for non-invasive, real-time monitoring of oxygenation conditions within biochips with cells and tissue samples. Such systems are more flexible and versatile than fibre-optic O$_2$ micro-sensors or solid-state coatings embedded in the biochips.

With an appropriate set-up, ecO$_2$ probes can be used in conjunction with commercial biochips to conduct simple O$_2$ assays and OCR measurements, an approach which is particularly useful when measuring with microscopic samples containing delicate cells preparations. They can also be used to optimise assay conditions and conduct the O$_2$ assays in a convenient and reproducible manner on commercial fluorescent readers. For example, Ibidi μ-slide microchambers (Fig. 2.6a and [50]) have been designed for optical microscopy imaging of both suspension and adherent cells that can be loaded manually with a micropipette. Such biochips with surfaces treated for cell attachment can also be used in pro-longed experiments with adherent cells.

Thus, using 6-channel Ibidi μ-slides, oxygenation conditions and cell behaviour under resting or re-perfusion conditions, were analysed using the MitoXpress$^{®}$ probe and measurements on a standard TR-F reader equipped with an adaptor holder for microscopy slides. The analysis of non-differentiated PC12 cells seeded at different densities inside the biochip, revealed a rapid deoxygenation of the chamber (Fig. 2.6b). Thus, at 60,000 cells per chamber (in 30 μL of medium) a level of 150 μM O$_2$ was reached after 2–3 h of monitoring. This is noteworthy as such densities are commonly used for DNA transfection while at higher cell densities anoxic conditions were quickly established. These data show that under the static conditions such as those used during cell attachment, cell seeded in air-saturated medium ($\sim$200 μM) can become deeply deoxygenated and experience hypoxia-induced physiological responses. From Fig. 2.6b, it is also evident that the rate of diffusion of ambient air O$_2$ through Ibidi μ-slide microchamber material is significant. At densities of 15,000 and 60,000 cells per chamber steady-state conditions were established after 2–3 h whereby cellular O$_2$ consumption is balanced by back diffusion. Therefore by changing the microchamber material, cell density, atmospheric pO$_2$ or flow conditions one can modulate cell oxygenation. These results also show that this measurement platform can be applied for rapid OCR measurements. Respiration profiles and such measurements tend to be faster and more sensitive than standard 96-well microplate measurements while also allowing repeat treatments by flushing the microchambers with fresh air-saturated medium, or drug.

2.2.6 Analysis of Microbial Cell Cultures

While OCR measurement is widely applied to the analysis of mammalian cells and organelles, there are also a wide variety of microbiological applications, both prokaryotic and eukaryotic. These are often based on rapid cellular growth and therefore require a slightly different approach: at low cell numbers the level of depletion of dissolved O$_2$ is not sufficient out strip back diffusion resulting in no signal change since the sample remains oxygenated. However, as bacteria replicate the level of O$_2$ consumption increases until a critical point is reached and the sample begins to deoxygenate rapidly causing a robust increase in the signal from ecO$_2$ probe. Figure 2.7a illustrates this type of data output whereby increasing *E.coli* seeing concentration results in earlier spike in probe signal. Such curves are amenable to an 'onset-time' data analysis approach where the time taken to reach a defined signal threshold is the output metric. Figure 2.7b illustrates the relationship between seeding concentration and such a metric, with nice linear plots in semi-logarithmic scale. Again, phosphorescence lifetime is a preferred readout parameter as it is more stable than intensity signals.

The simplicity and robust nature of the oxygenation-based growth profiles generated using phosphorescent O$_2$ probes has led to it being adapted for a variety of application. Examples include the enumeration of bacterial load in complex samples such as food homogenates, environmental and wastewater samples using microtiter plates (96 and 384-well), thereby avoiding the need for standard aerobic plate counting and multiple dilutions [20]. Figure 2.7c illustrates the data output of such analysis of ground beef with higher bacterial load reflected by an earlier onset-time. These onset times can then be converted to bacterial load values, using a pre-determined calibration. Figure 2.7d shows this conversion with load presented in CFU/mg and a 'flag' calculated based on defined acceptance criteria.

This measurements approach can also be used for the optimisation of culture conditions, for screening applications and determination of drug resistance. A dose response analysis is presented in Fig. 2.8 where *S.aureus* seeded at $\sim 1 \times 10^7$ cells/ml are exposed to increasing concentrations of antibiotic and O$_2$ depletion is measured kinetically. The respiration profile for all test wells is presented and, as

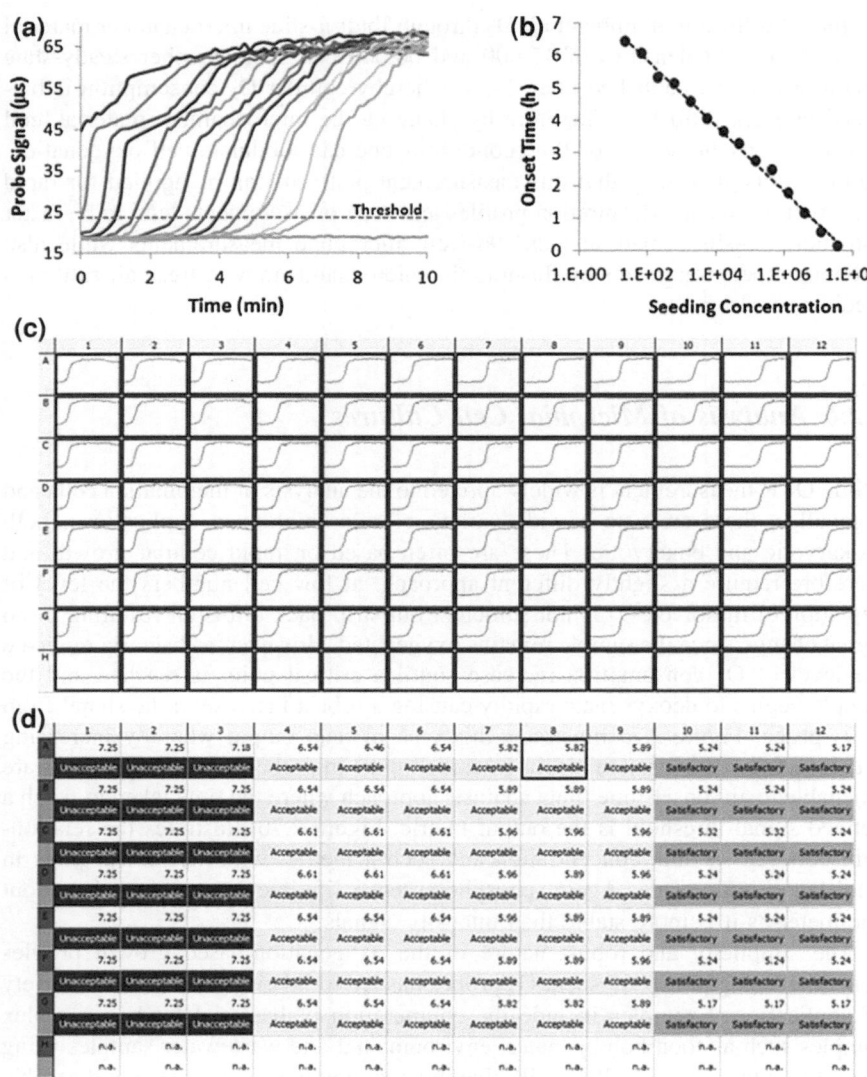

Fig. 2.7 Assessing bacterial respiration showing O₂-based E.coli growth curves (**a**), the relationship between seeding concentration and onset-time (**b**), the data output from bacterial load assessment showing the profiles measured in each well (**c**) and the calculated bacterial load in CFU/mg and a 'flag' based on this load (**d**)

above, the time at which probe signal increases indicates the degree to which growth has been inhibited with flat lines indicating complete inhibition. Onset times can, subsequently, be calculated for the generation of IC₅₀ and MIC values.

Such an approach can also be applied to the analysis of yeast metabolism where a longer doubling time means that high cell numbers and short measurement times

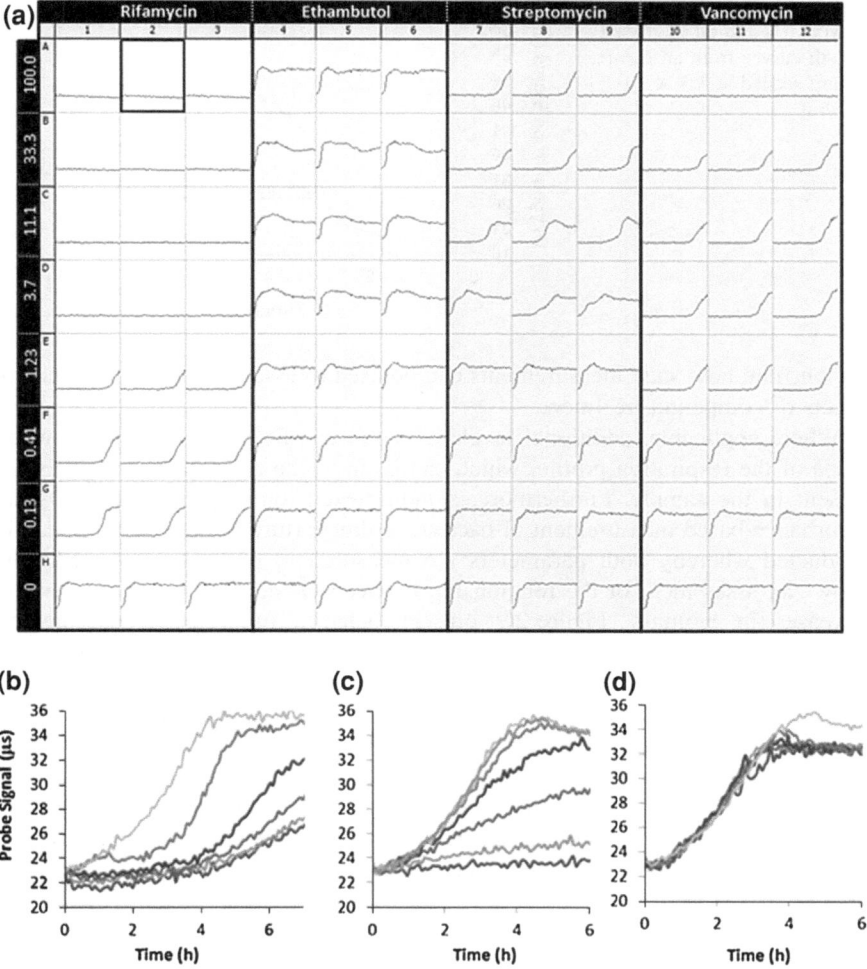

Fig. 2.8 a *S.aureus* seeded at ∼1 × 10⁷ cells/ml in EB broth, exposed to increasing concentrations of the indicated antibiotic and measured kinetically at 37 °C. The drug tested is indicated above the relevant wells with the treatment concentration to the left. **b–d** O₂ consumption profiles from *C.albicans* (∼3 × 10⁵ cells/ml) treated with increasing concentrations Ant A (from 30 μM), Amphotericin (from 16 μg/ml) and Fluconazone (from 65 μg/ml) in RPMI medium

allows the assessment of immediate effects on cell metabolism while using lower cell numbers and extended measurement times facilitates analysis of effect on cell growth and metabolism. Figure 2.8a presents a series of *C.albicans* dose responses. The ETC inhibitor Ant A (Fig. 2.8b) and the polyene antifungal Amphotericin B (Fig. 2.8c) cause an immediate and dose-dependent decreases in O₂ consumption while the triazole antifungal Fluconazole (Fig. 2.8d) caused no appreciable decrease in OCR. These observations correlate with mode of drug action and

Fig. 2.9 Comparison
between O_2 and OD_{600} based
growth curves from an *E.coli*
culture seeded at 1.3×10^5
cells/ml

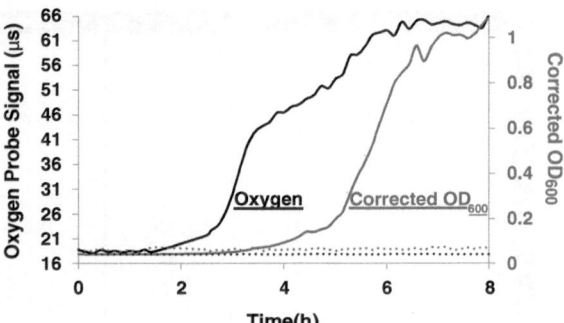

demonstrate how such measurements can be used to assess the specific metabolic effects of compound treatment.

Where applicable, additional levels of detail can be gleaned from the whole shape of the respiration profile, which can point to the type of microorganism(s) present in the sample. Furthermore, a multiplexed luminescence-based O_2 and absorbance-based measurement of bacterial cultures (turbidimetry, OD_{600}) can be conducted whereby both parameters are measured in the same test well. This allows an assessment of the relationship between O_2 dependent metabolism and increases in biomass. Figure 2.9 presents such a multiplexed measurement whereby *E.coli* seeded at $\sim 1 \times 10^5$ cells/ml are monitored with O_2 depletion occurring significantly earlier than measurable increased in OD_{600}. Such measurement can also be used to assess the contribution of aerobic metabolism to the energy required for growth, but such a multiplexed approach can only be applied under certain conditions as stable OD_{600} growth curves can sometimes be difficult to generate particularly in opaque or coloured growth media. O_2 measurements are much more flexible and robust in this regard, particularly when using the lifetime-based measurement modality and data analysis approach.

2.2.7 Respiration of Small Organisms and Model Animals

While the majority of applications of optical O_2 micro-respirometry and OCR measurement are cell-based, the method has also been adapted for the monitoring of small aquatic organisms including *Daphnia magna*, *Artemia salina*, copepod *Tigriopus*, zebrafish *Danio rerio* [3, 4]. These model organisms can be used to perform rapid, high-throughput biological testing of potentially hazardous chemicals and environmental samples, as well as basic biological and physiological studies. In such assays, test organisms are exposed to a particular condition (development stage, environment, potential toxicant or stress) and then allowed to respire in a sealed compartment in the presence of an ecO₂ probe. The resultant O_2 depletion causes an increase in probe phosphorescence over time, reflecting the metabolic activity of the organism. The appropriate measurement platform is selected from the available options (Chap. 1, Fig. 1.5), based on the respiration activity and size of test organism, and specific application requirements.

For example, this method was applied to the measurement of individual *Daphnia magna* after 24 and 48 h toxicant exposures, using low-volume sealable 96-well plates and MitoXpress® probe [51]. The respirometric measurements performed with reference toxicants including $K_2Cr_2O_7$, sodium lauryl sulphate and heavy metals, showed good agreement with the established *Daphnia* test based on mortality assessment while the ability to analyse sub-lethal effects and complex samples such as industrial effluents, facilitates the generation of dose–response relationships and EC_{50} values. Representative profiles of *Daphnia magna* respiration are shown in Fig. 2.10. Zebrafish *Danio rerio*, another useful model for genetic manipulation, toxicological studies and environmental monitoring, was also applied for toxicological assessment by measuring respiration of individual embryos (48 h after hatching) [52].

Organisms such as *Caenorhabditis elegans* have also been measured. This is an interesting model organism as it is a non-parasitic, multicellular metazoan with a 3-day lifecycle and is commonly used in developmental biology, behavior, anatomy and genetics studies offering relevant endpoints such as mortality, life span, behavior/movement, feeding, growth and reproduction. This invertebrate organism easily maintained under laboratory conditions is an attractive model for toxicology studies as it is sensitive to a wide range of toxicants, including heavy metals, organic phosphates and pesticides. Such studies have established *C. elegans* as a powerful model for rapid testing of the toxicity of soil and water samples as well as pharmaceutical compounds [53].

Another system for toxicological assessment uses panels of test organisms which include prokaryotic (*E. coli*, *V. fischeri*) and eukaryotic (Jurkat) cells, invertebrate (*Artemia salina*) and vertebrate (*Danio rerio*) organisms. Convenient assay set-up allows parallel assessment of up to 96 samples or data points in ~ 2 h, and the generation of dose and time-dependent responses in both standard and low-volume 96-well plates. The methodology was demonstrated with discrete chemical classes including heavy metal ions, PAHs and pesticides, chemical mixtures and complex environmental samples such as wastewater from a WWTP. Compared to the single organism testing, this system provides more detailed information and allows profiling of different toxicants on the basis of the pattern of their response. Representative results and patterns of toxicity are shown in Fig. 2.11. The panel of organisms can also be modified or extended.

Such toxicological screening systems provide high sensitivity, sample throughput and information content, flexibility and general robustness. They allows ranking and profiling of samples and compare favourably with the established methods such as MicroTox®, Daphnia Test and mortality tests with animal models. They are well suited for large-scale monitoring programmes such as the US Clean Water Act, EU Water Framework Directive and EU REACH (Registration, Evaluation and Assessment of Chemicals) programme.

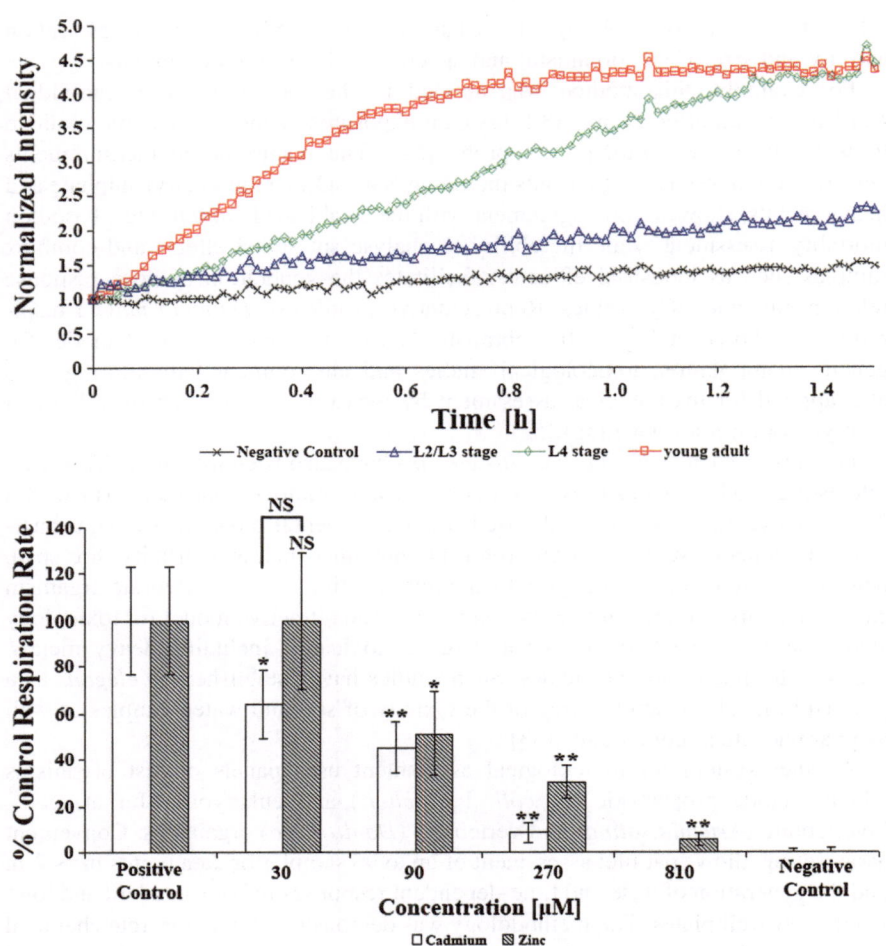

Fig. 2.10 Respiration profiles of individual *C. elegans* at different age (*top*) and the effects of their 24 h exposure to different concentration of heavy metal ions (Zn and Cd, as indicated). Measured in low-volume sealable 96-well microplates at room temperature. Reproduced from [53], with permission of Wiley

2.2.8 Enzymatic Assays

O$_2$-consuming enzymes play an important role in many biochemical pathways. They also have diagnostic value and screening potential, and therefore have become the focus of attention both as drug targets and as players in drug metabolism. Common examples include the monoamine oxidase (MAO), cytochrome P450 (CYP450) and cyclooxygenase (COX) families, glucose oxidase and

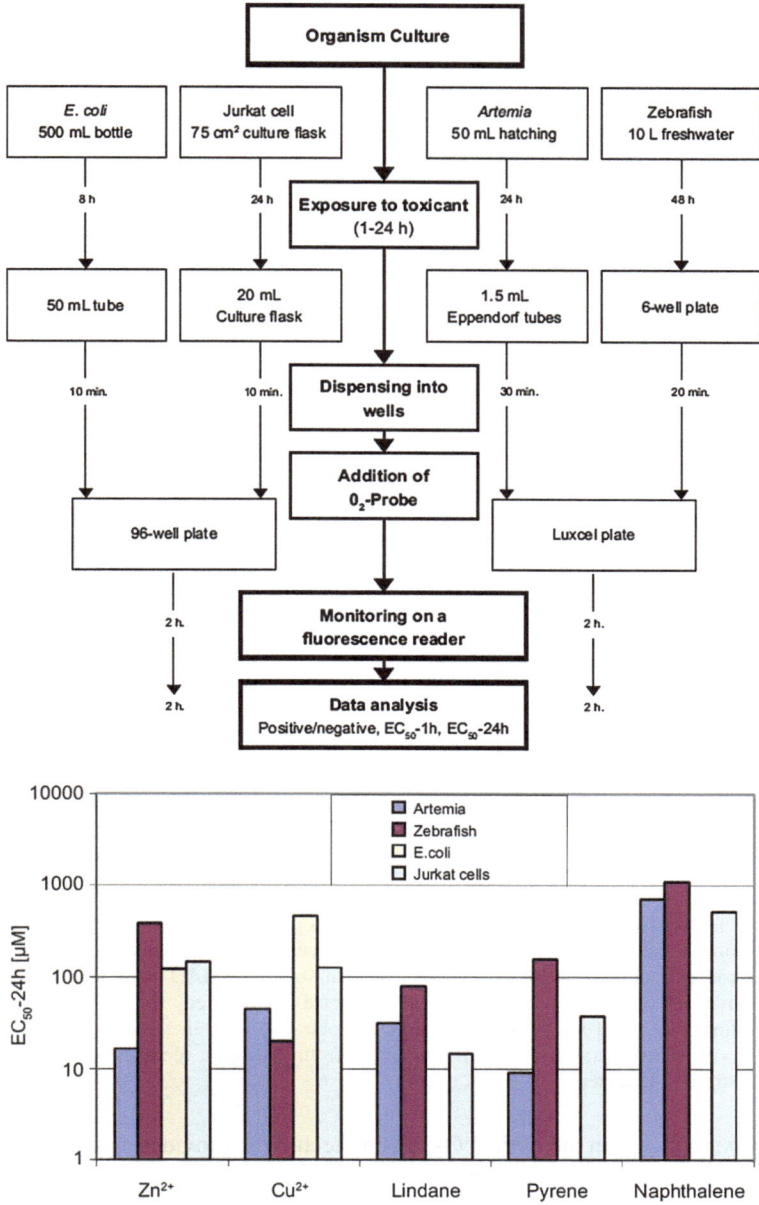

Fig. 2.11 Scheme of toxicological assessment of samples using panels of test organisms and O₂ respirometry (*top*) and representative results (*bottom*). Reproduced from [52] with permission of Wiley

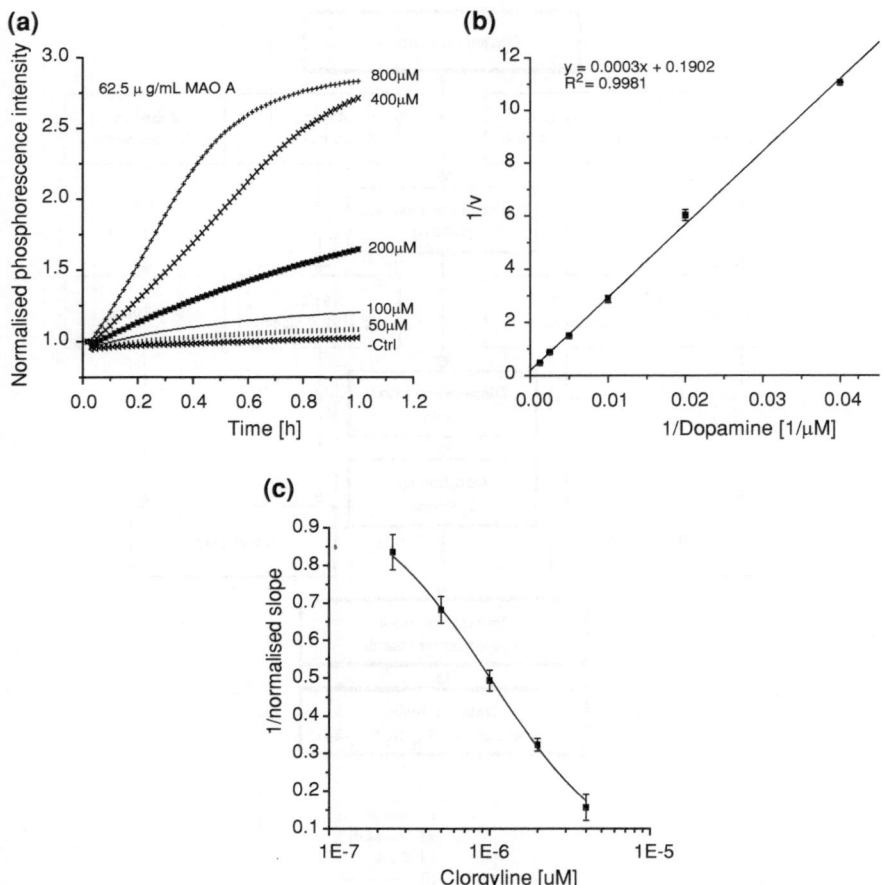

Fig. 2.12 a Normalised intensity profiles for 62.5 µg/ml MAO-A enzyme at different concentrations of dopamine; **b** The relationship between reciprocal reaction rate (*slope*) and substrate concentration (derived from A) is linear. **c** Inhibition of MAO-A activity (0.125 mg/ml and 0.4 mM of dopamine) by clorgyline: inversed normalised slopes versus inhibitor concentration. Measured with 5 µM MitoXpress probe at 37 °C, in three replicates for each point ($n = 3$, error bars are shown). Adapted from [6] with permission of Analytical Biochemistry and Elsevier

lactate oxidase. Again the optical O$_2$ sensing technique provides convenient means for monitoring of activity and inhibition of a particular enzyme quantification of their substrates and inhibitors, control of production and characterisation of recombinant enzymes (oxidases). Similar to the mitochondria and cell-based applications described above, such enzymatic assays can be conducted in a convenient 96/384-well plate format using simple mix-and-measure procedure and measurement on a fluorescent plate reader of samples sealed under oil, similar to the assays with isolated mitochondria (see Sect. 2.2.1).

The capillary-based LightCycler system was also shown to be efficient in such applications, providing assay miniaturisation, better sensitivity and reproducibility compared to microplate assays. Using the phosphorescence-based molecular and nanoparticle probes solid-state sensor coatings, this platform was demonstrated with the analysis of important oxygenases (MAO, COX and CYP450 families) and their specific inhibitors [6]. Representative O$_2$ consumption profiles and their transformation into calibration curves for the quantification of the substrate and inhibitors are shown in Fig. 2.12. Sensitive detection of cholinesterase inhibitors in a coupled enzymatic system of acetylcholinesterase—choline oxidase was also demonstrated using optical O$_2$ detection and the LightCycler® system [54].

Besides the above examples and case studies, a number of other specimens, including plant mitochondria, seeds (analysis of germination), different types of microbial and mammalian cells, have been analysed and used in various mechanistic studies on cell metabolism, mitochondrial function and signalling, in conjunction with the phosphorescent ecO$_2$ probes and plate reader analysis. All these demonstrate the versatility of this sensor chemistry and measurement methodology and the large number of analytical tasks they can perform.

2.3 Bioassays Performed with icO$_2$ Probes

It is known that oxygenation of live tissue is significantly lower than dissolved O$_2$ concentration in air-saturated solution (~ 200 μM or 158 mmHg at normal atmospheric pressure, 20.9 % O$_2$, 37 °C). Thus, measured in arteries, brain and retina O$_2$ levels were found to be ~ 97.5, 1–40 and 2–5 mmHg, respectively [55, 56]. Upon pathological conditions or disease states O$_2$ levels in the cells may further decrease, leading to energy stress, activation of adaptive responses and, ultimately, cell survival or death. Therefore, it is difficult to overestimate the importance of precise monitoring of the cell oxygenation as a key metabolic parameter, which informs on cell functioning and helps to predict cell fate. Pt-porphyrin-based icO$_2$ probes are well suited to quantitative real-time monitoring of O$_2$ in various biological samples, from cell populations to sophisticated 3D specimens and O$_2$ mapping in complex heterogeneous samples (as outlined in Chap. 3).

2.3.1 Relationship Between Respiratory Activity, Sample Parameters and Oxygenation

Mathematical modelling facilitates a better understanding of the various biophysical processes taking place in biological samples [57, 58] and helps with the comparison of experimental data and theoretical models. A limitation, however, is that available experimental techniques often have limited capacity to explicitly

Fig. 2.13 Physical model of
a biological sample
containing a monolayer of
respiring cells in growth
medium exposed to
atmospheric O_2

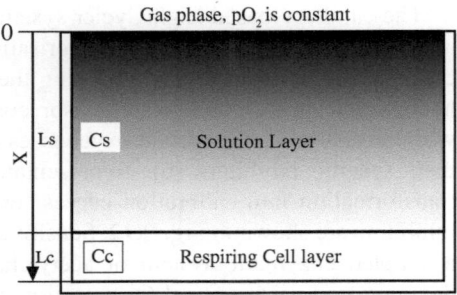

confirm the models in question. New techniques are therefore required to address
this gap and the O_2 measurements described here contribute to this. Since respiration of cells and tissues has a complex dependence on O_2 concentration, a
precise analytical description of O_2 dynamics in different parts of a 3D sample is
difficult. However, for simpler models often used in in vitro studies, such as
individual cells suspensions (spherical model), monolayers of adherent cells (2D)
or spheroids and scaffolds (3D), mathematical description of localized O_2 gradients and O_2 mapping, can be realised in relatively simple mathematical terms [59].

Thus, for a simple planar model of adherent respiring cells maintained in a
vessel with headspace under static culturing conditions (e.g. samples in microplate
wells—Fig. 2.13), O_2 concentration in the solution layer (Cs) and within the cell
layer (Cc) can be described by simple mathematical equations. Using a number of
valid assumptions, one can work out that the profiles of Cs and Cc within the
sample obey the linear and quadratic functions of the distance from the headspace
which serves as an O_2 reservoir, and the thickness of the Ls and Lc layers [59]:

$$Cs = C_0 - \frac{Lc}{Ds}k(X + Ls) \tag{2.1}$$

$$Cc = \frac{k}{2Dc}X^2 - k\frac{Lc}{Dc}X + H\left(C_0 - \frac{LcLs}{Ds}k\right) \tag{2.2}$$

In these equations, C_0 is O_2 concentration at the gas/solution interface ($X = 0$) at
given O_2 and temperature; Dc and Ds–O_2 diffusion coefficients in solution and cell
layers; H–O_2 partition coefficient at cells/solution interface ($Cc|_{X=Ls} = HCs|_{X=Ls}$);
X—distance from the interface and k—specific respiratory activity of the cell layer.
The model suggests that actively respiring cells can generate local O_2 gradient in the
sample, and this can be assessed by means of:

An ecO_2 probe which informs on the Cs in the medium. When measured on a
plate reader through the sample, it produces Cs values averaged across the whole
depth of medium layer above the cells;

An icO_2 probe which informs on Cc within cell monolayer. Due to the small
thickness of this layer (~ 10 μm), the O_2 gradient across it is deemed very small so
that oxygenation of the cells is relatively uniform and corresponds to the measured
Cc value. Hence, icO_2 probes enable accurate measurement of oxygenation of cell

layers and through this parameter monitoring of sustained or transient changes in respiration, associated with cell metabolic activity, the levels of atmospheric pO_2 or drug action.

2.3.2 Cell Loading and Optical Measurements with icO_2 Probes

For probes developed specifically for intracellular use, loading usually involves addition to complete growth medium at concentrations optimised in separate experiments or provided by the vendor, and incubating the cells under standard culturing conditions for several hours (3–16 h or shorter) in a CO_2 incubator at 37 °C. After washing and change to fresh medium, the plate with cells is ready for measurements and treatments.

Although cell-permeable probes enable the measurement of icO_2 in a simple mix-and-measure format, a number of factors have to be considered. For reliable and accurate lifetime-based O_2 sensing, the phosphorescent signals produced by samples of loaded cells should be enough high to ensure signal-to-blank ratio (S:B) of at least 10. The intensity signals, in turn, are determined by the following parameters:

(1) Probe loading concentration and incubation time. With the example of NanO2 probe typical dependences can be seen in Fig. 2.14, which look almost linear without sign of saturation. This probe even when used at low very doses (2–5 µg/ml) produces very high signals in different cell types, which may exceed 10^5–10^6 cps (counts/photons per second) and even cause detector saturation at high concentrations. However, for other icO_2 probes loading efficiency and kinetics may be very different and acceptable levels of signals for reliable sensing of icO_2 may not so easy to reach. For the high-sensitivity TR-F reader Victor[4] in our lab the acceptable signal threshold was set at 30,000 cps and blanks were typically <1,000 cps. But again for another instrument the threshold and blanks may be quite different due to different instrument performance or intensity scale used (luminescence spectroscopy operates with arbitrary intensity units).

(2) The density or number of cells in the sample. Generally, the more cells in the sample the higher the phosphorescent signals after loading. At very low cell numbers the signals can drop below the acceptable threshold which will result in increased error in lifetime and O_2 determination. At low cell density local O_2 gradients vanish due to low levels of respiration (see below), while very high cell numbers can cause complete deoxygenation, hypoxic shock and death of the cells.

(3) Atmospheric O_2 levels and oxygenation of the medium and cell monolayer. Given that phosphorescence intensity signals are inversely related to O_2 concentration (see Eq. 1.1) and that the O_2 sensing method provides best performance and resolution at zero O_2 [12], measurements under low O_2

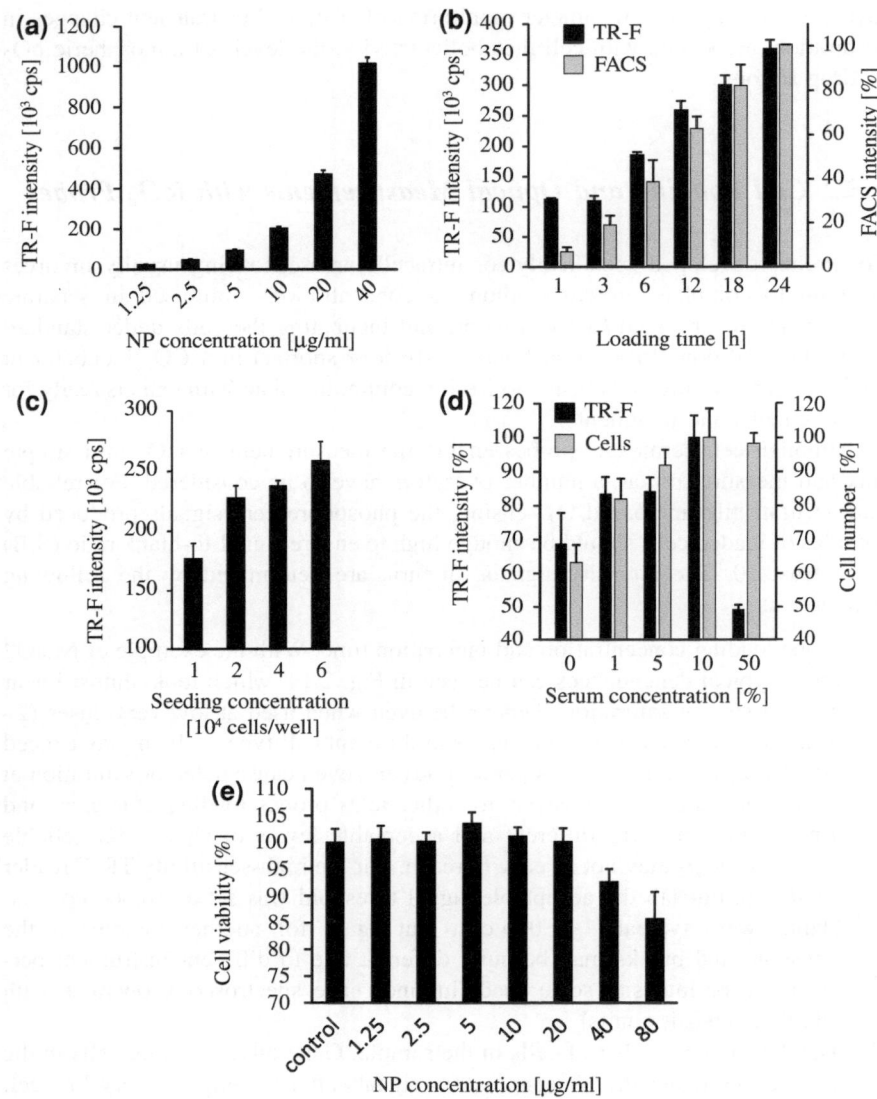

Fig. 2.14 Loading of NanO2 probe in MEF cells in DMEM medium, 37 °C. **a–d** The dependence of TR-F intensity signals on the probe concentration (**a**, for 12 h loading), loading time (**b**, for 10 μg/mL probe), cell seeding concentration (**c**) and serum content in the medium (**d**). **e**. The effects of probe on cell viability (15 h loading). Adapted with permission from [10]. Copyright 2011, American Chemical Society

atmosphere or cellular hypoxia can be performed with reduced cell numbers or lower probe loading, as compared to normoxic (ambient air) conditions.

(4) Medium composition, mainly serum content and the presence of various additives (e.g. transfection reagents or compounds that have an effect on cell membrane and endocytosis processes.

(5) Toxicological impact of the probe on cells which may result in cell death and loss of integrity, thus releasing the icO$_2$ probe into bulk medium. These effects have to be carefully analysed at the start of each study with a new model and monitored throughout the work.

2.3.3 Probe Calibration

Reliable calibration is key to successful and accurate sensing of O$_2$ by phosphorescence quenching. Particularly when performing O$_2$ calibrations of icO$_2$ probes, it is necessary to consider all the main factors that can potentially influence measured lifetime values and the O$_2$ concentrations calculated from these values. Although phosphorescence lifetime is the intrinsic parameter of the probe, its measurement may be affected by: (i) instrumental factors which include measurement errors and systematic inaccuracy originating from the instrument hardware or software; (ii) changes in probe micro-environment; (iii) calibration set-up and operator errors; (iv) errors in converting lifetime values into O$_2$ concentration.

It is, therefore, recommended to perform the calibration in conditions that closely resemble those used in the biological experiments, including the instrument, probe localisation, optical alignment, intensity signals, data acquisition and processing algorithms, O$_2$ concentration range. Although some icO$_2$ probes (e.g. nanoparticle based) are shielded from quenching interferences and effects of changing microenvironment, variability in lifetime readings can be still observed when the same O$_2$ probe is measured on different instruments or in different cell types. Some of the commercial instruments (claimed to be suitable for sensitive TR-F measurements), in fact, show a large variability and inaccuracy of measured lifetime and a pronounced intensity dependence. Such instruments are therefore not suitable for quantitative O$_2$ sensing experiments.

To precisely control O$_2$ in the microplate compartment, the instrument should have a built-in O$_2$ controller (e.g. FLUOstar Omega, BMG Labtech). Standard TR-F readers such as Victor, PerkinElmer can be placed in a hypoxia chamber with O$_2$ control (e.g. Coy Laboratory Products, MI). When calibrations are performed with standard gas mixtures in a hypoxia chamber, correction for variations in atmospheric pressure should be made. For example, at normal atmospheric pressure (1 Bar or 760 mmHg), 10.0 % v/v O$_2$ in hypoxia chamber corresponds to partial pressure of 10.0 kPa O$_2$ or dissolved concentration 100 μM.

Based on our experience with different icO$_2$ probes, the following calibration procedure is recommended:

(1) Seed test cells in multiple wells of a microplate and culture them to medium/high confluence;

(2) Load the cells with probe in standard growth medium, to achieve sufficiently high TR-F intensity signals in ambient air environment that ensure reliable lifetime measurements (see above typical thresholds). Include wells with unloaded cells as blanks;

(3) Add Ant A to the cells (5–10 μM) to inhibit cellular respiration and bring cellular O$_2$ to the levels in bulk medium. This requires ∼15 min incubation.

(4) Insert the plate with cells in the reader pre-equilibrated at the desired temperature (normally 37 °C) and atmospheric O$_2$ (can start with ambient, 20.9 %).

(5) Initiate ratiometric TR-F measurements in kinetic mode monitoring probe signal in assay wells every 5–10 min for about 20–40 min until stable intensity and lifetime readings are achieved. This corresponds to gas and temperature equilibration of samples in the well.

(6) Change atmospheric O$_2$-level and repeat the monitoring cycle. Make this step at several O$_2$ levels covering the range of interest, for example, 20.9 %, 15.0 %, 10.0 %, 5.0 %, 3.0 % and 1.0 % O$_2$.

(7) In the end of the protocol add glucose oxidase and β-D-glucose to the samples (50 μg/ml and 10 mM, respectively) and measure lifetimes as above for fully deoxygenated samples (zero O$_2$ point).

(8) Calculate lifetimes for the plateau region for each O$_2$ level. For several wells and measurement points determine average lifetime values and standard deviations.

(9) Use these data points to plot a calibration graph. Perform best fit of the data points to determine the calibration function (mathematical equation): $[O_2] = f(\tau)$.

Representative experimental data, O$_2$ calibration graph and analytical function derived from them, are shown in Fig. 2.15. Figure 2.15a also shows that in the respiring MEF cells measured lifetime values are significantly higher and O$_2$ levels lower than the standards, so they cannot be used for the calibration. During the prolonged calibration measurement gradual evaporation of medium should be compensated by adding water to the samples when the level drops below 50 %.

2.3.4 Control of Oxygenation in Cultures of Adherent Cells

The model presented in Sect. 2.3.1 and Fig. 2.13 suggests that actively respiring cells can partly or fully deoxygenate themselves and generate O$_2$ gradients that propagate across the sample. Deoxygenation due to intrinsic respiration (cells are acting as O$_2$ sinks) is particularly important for actively respiring cells and cells cultured under reduced atmospheric O$_2$ levels. Figure 2.16 shows lifetime and icO$_2$ profiles monitored on a TR-F reader in the resting MEFs seeded at different density on 96-well plate, grown for 12 h at 20.9 % O$_2$ and then transferred to 3 % O$_2$. The initial portion of the curve (0–45 min) reflects the establishment of new

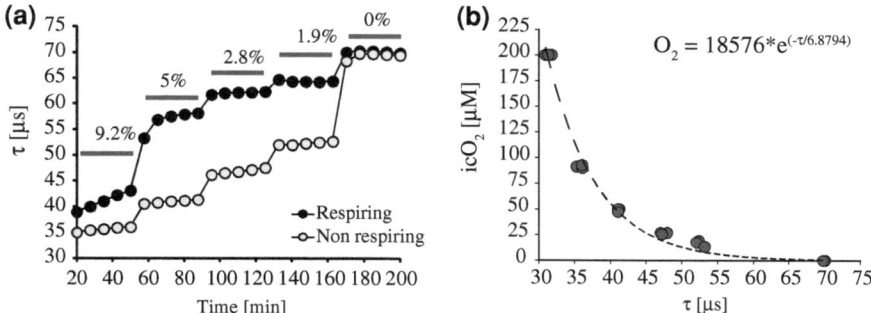

Fig. 2.15 Calibration of the NanO2 probe in MEF cells generated on Victor4 TR-F reader and hypoxia chamber (glove box, Coy Laboratory Products) at 37 °C. A. Experimental steps of the probe calibration between 9.2 % and 0 % atmospheric O$_2$. Oxygenation of respiring cells is also shown. B. Analytical equation for calculating O$_2$ concentration (µM), produced by fitting the data points (3 independent experiments) with exponential function

steady state between O$_2$ consumption and supply, which is characteristic for particular atmospheric O$_2$ level and respiratory activity of the sample.

These results show that respiring cells in static culture do deoxygenate themselves and can easily become anoxic when O$_2$ availability decreases (hypoxia). Inhibition of respiration by Ant A eliminates this effect while the cells still remain viable (ATP levels unaffected) due to the compensatory activation of glycolysis outlined above. At 20.9 % atmospheric O$_2$ oxygenation of MEFs seeded at the same density varies between ~180 µM (10^4 cells) and ~140 µM (5×10^4 cells). Similar profiles of cellular O$_2$ are expected for tissue culture flasks maintained under static conditions at which steady-state O$_2$ gradients are also generated. These O$_2$ gradients can be disrupted by agitation or shaking the flasks and microplates.

The critical observation to draw from these data is that precise O$_2$ levels set in the hypoxia chamber (e.g. 0.5 %, 1.0 % or 5.0 %) do not translate into the same O$_2$ levels within the cells. Such a relationship only holds at very low cell densities. Therefore, careful control of experimental conditions is required to avoid extreme fluctuations in cellular oxygenation.

2.3.5 Effects of Cell Metabolic Activity on icO$_2$

Besides atmospheric O$_2$, cellular respiratory activity (parameter k in the model and Eqs. 2.1, 2.2) also has a large effect on icO$_2$, and this can be explored to study changes in cell metabolism, compare different cell types or mutant cells, apply drug treatments and investigate mechanism of action. Thus, Fig. 2.17 shows that Krebs cycle deficient MEFs (see Sect. 2.2.4 and [41]), when transferred from normoxia to moderate hypoxia (8 % O$_2$), have a much higher deoxygenation rate than WT MEFs. Parallel measurement of cells sealed under oil with an ecO$_2$ probe

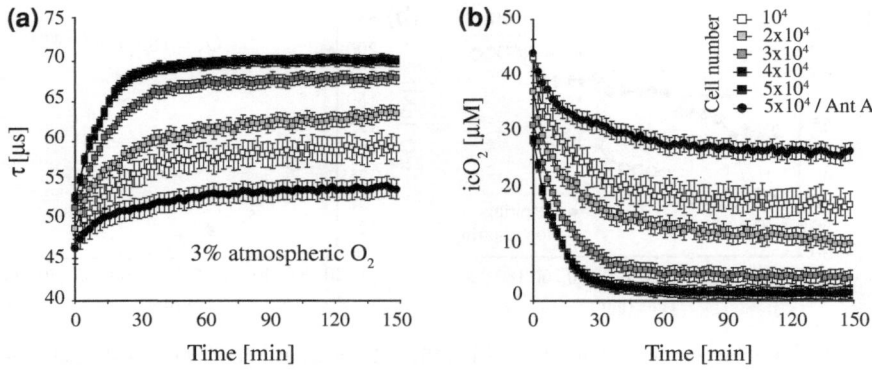

Fig. 2.16 Measured lifetime (**a**) and calculated O₂ (**b**) profiles for resting MEFs cultured in DMEM medium at different seeding densities (*indicated*). Experiment was performed using NanO2 probe under 3 % atmospheric O₂ at 37 °C

confirmed that OCR in KO cells was much lower and did not increase in the presence of FCCP.

Similarly, treatment of the cells with drugs that target cell metabolism can have a profound influence: an increase in icO_2 (or reduction in probe lifetime) points to the inhibition of cell respiration, whereas reduction in icO_2—to the activating/ uncoupling effect. For example, treatment of PC12 cells with a well-known inhibitor of V-ATPase and K^+ ionophore Bafilomycin A (Baf) [60], [61] revealed its pronounced effect as a mitochondrial uncoupler [15]. This was seen as a reduction of icO_2 even under ambient atmosphere (20.9 % O_2) which occurs in a dose and time-dependent manner.

In MEFs, the respiratory response to Baf was also dose dependent and in normoxia became significant at ≥ 0.25 μM Baf (Fig. 2.18a) without significant effect on cellular ATP levels. Since relative deoxygenation of respiring cells increases with decreased O_2 availability [59], in agreement with this rule at 6 % atmospheric O_2 the effect of Baf became more pronounced and detectable already at concentration 0.06 μM (Fig. 2.18b), while 0.2–0.5 μM Baf caused a sustained decrease in icO_2 down to 1–5 μM. An increase in cell respiration was linked to significant changes in mitochondrial function (Fig. 2.19). Thus, average intensity of TMRM (probe for mitochondrial membrane potential, ΔΨm) was partially decreased and the analysis of co-localization of TMRM and $_{mito}$Case12 (non-leaking mitochondria-targeted Ca^{2+}-biosensor) revealed continuous flickering of ΔΨm. The behaviour of MEF cells and effects caused by Baf were similar to what was previously found in PC12 cells and described in greater detail elsewhere [15].

Cell deoxygenation induced by Baf can be abolished by the addition of Ant A. After such double treatment cellular ΔΨm and ΔpH$_m$ rapidly disappeared, since complex III could not contribute to their restoration, while FoF1 ATP-synthase consumed glycolytic ATP to pump protons outside the matrix. Inhibition of

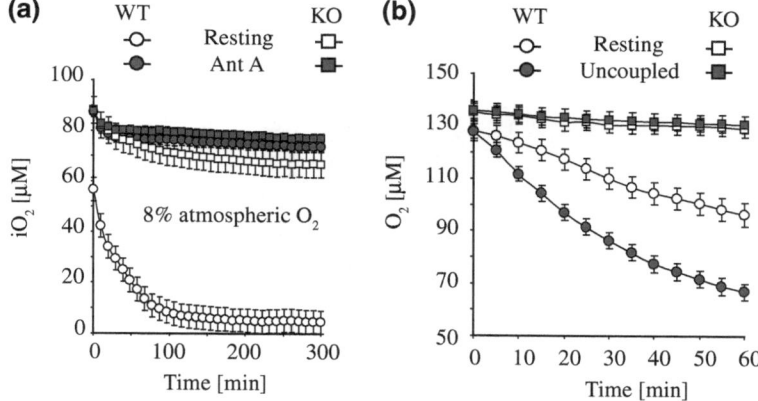

Fig. 2.17 The effects of respiration activity and metabolic status of the WT and FH deficient MEF cells on icO₂ (**a**) and their correlation with the OCR measured with an extracellular probe under oil seal (**b**)

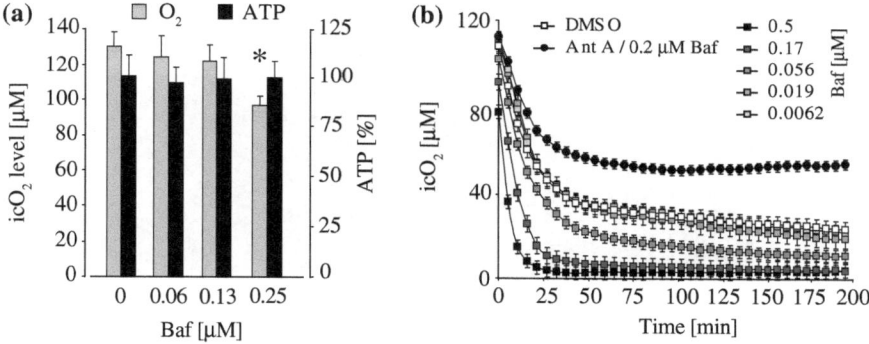

Fig. 2.18 Uncoupling effect of Baf on icO₂ and ATP levels in MEF cells. **a** At normoxia (21 % O₂) treatment with 0.25 μM Baf causes significant reduction in icO₂ and does not affect cellular ATP levels. **b** In hypoxia (6 % atmospheric O₂) the effect of Baf on icO₂ levels is seen at >0.056 μM concentrations. Deep deoxygenation of MEFs lasts for several hours. In all cells treated with Ant A (10 μM) the O₂ does not differ from atmospheric level due to inhibition of the mitochondrial complex III. Asterisks indicate significant difference from mock control (DMSO)

respiration is, therefore, extremely stressful for the cells. However, physiological feedback mechanisms that exist in the cell, such as nitric oxide (NO) signalling, can regulate the respiratory responses and cell oxygenation. In the presence of NO the cells cannot deoxygenate themselves below a certain level. This is because NO effectively competes with O₂ for cytochrome c oxidase (mitochondrial complex IV) and inhibits its activity and at reduced O₂ availability decreases cell respiration to a certain level in a manner dependent on cellular O₂ and NO concentrations [59]. Figure 2.20a demonstrates that in the presence of NO donor DETA-NONOate (1 mM), which maintains ∼0.5 μM NO in the medium, respiration was

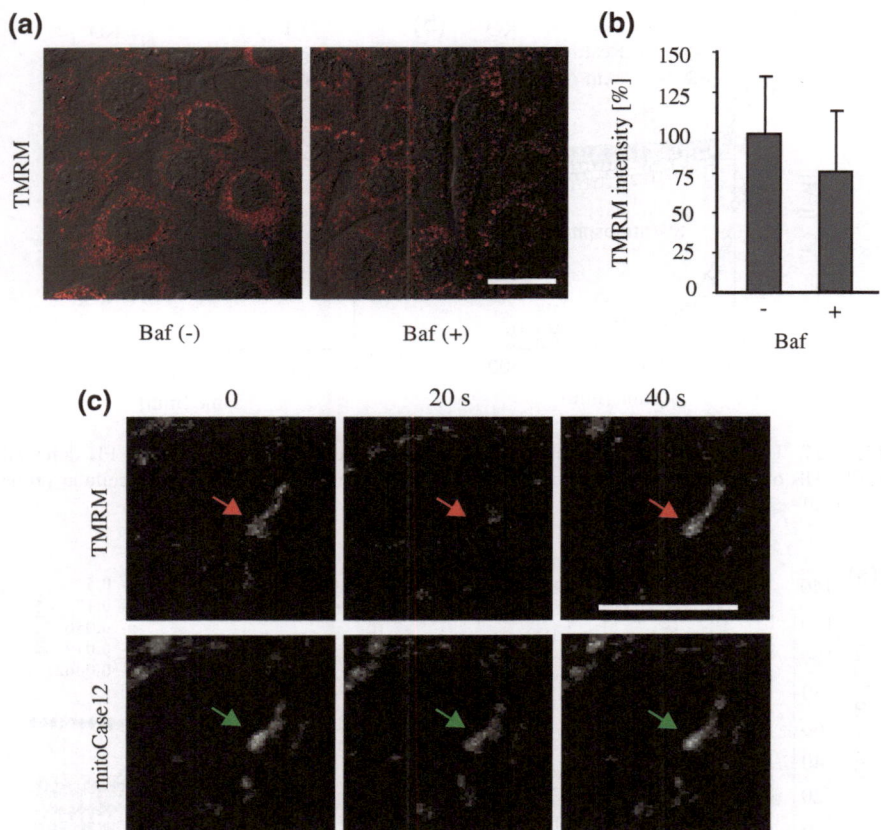

Fig. 2.19 Effect of Baf treatment on the $\Delta\Psi$m polarisation in MEFs. **a** Baf treatment (0.25 μM for 45–60 min) causes slight depolarisation of the $\Delta\Psi$m, as monitored using TMRM. **b** Semi-quantitative analysis of TMRM fluorescence intensity is shown in (**a**). **c** Baf induces random flickering of the $\Delta\Psi$m with an average frequency of 20 s (*red arrow*), as seen in the single mitochondrion visualised with a mitochondrial Ca^{2+} sensor mitoCase12 (*green arrow*). Scale bar represents 20 μm

inhibited when O_2 levels drop to about 20–25 μM. Cellular ATP levels remains steady and independent of the treatment with Baf and DETA-NONOate (Fig. 2.20b). At ~4.5 % atmospheric O_2, such inhibition of respiration occurs regardless of Baf treatment, demonstrating universality of NO signaling.

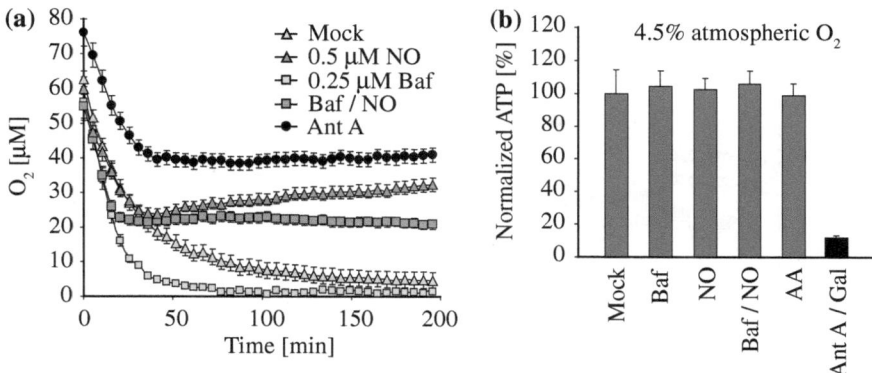

Fig. 2.20 Effect of NO on the respiratory response to Baf of differentiated PC12 cells monitored at 4.5 % O₂. **a** Cell deoxygenation profiles. **b** The levels of cellular ATP

2.3.6 Multiple Treatments with Drugs and Monitoring of Transient Responses of Cells

As icO₂ concentration is a function of O₂ availability, conditions of mass exchange and cell respiratory activity, its levels may be affected significantly by rapid metabolic stimulation whereby a transient imbalance between O₂ supply and demand leads to a rapid local decrease in icO₂ ('overshoot effect' [59]). Even a short, transient increase in respiration may lead to sustained deoxygenation, imposing an energy stress which can in turn trigger cell adaptation to hypoxia. The latter can be seen, for example, as a shift in energy production from OxPhos to glycolysis, activation of hypoxia inducible factor (HIF) pathway and change in gene expression profile [59]. When measured in icO₂ or lifetime scale, the shape of cellular responses to metabolic stimulation is determined by the type of cells, their physiological environment (growth medium, supplements, etc.), mode of drug action and diffusion characteristics of the measurement platform. By changing experimental conditions such as, the volume of medium, atmospheric O₂ or temperature, one can modulate the observed response and extract specific mechanistic information about cellular function. For example, rapid depletion of extracellular Ca^{2+} (eCa^{2+}) with EGTA was shown to cause transient elevation of respiration in PC12 cells, through Na^+ influx and a subsequent increase in mitochondrial Na^+/Ca^{2+} and H^+/Na^+ exchange [17]. EGTA also slightly decreased the respiratory response to subsequent plasma membrane depolarisation by KCl [16, 17]. Based on this knowledge, we investigated how the depletion of eCa^{2+} affects the response to FCCP uncoupling.

PC12 cells pre-incubated in glucose(+) and galactose(+) media for 3 h and were treated sequentially with 5 mM EGTA and 1 μM FCCP (Fig. 2.21). Glucose(+) cells exhibited higher resting icO₂ than galactose(+) cells, and a lower response to FCCP which was slightly inhibited by eCa^{2+} depletion. In galactose(+) medium for

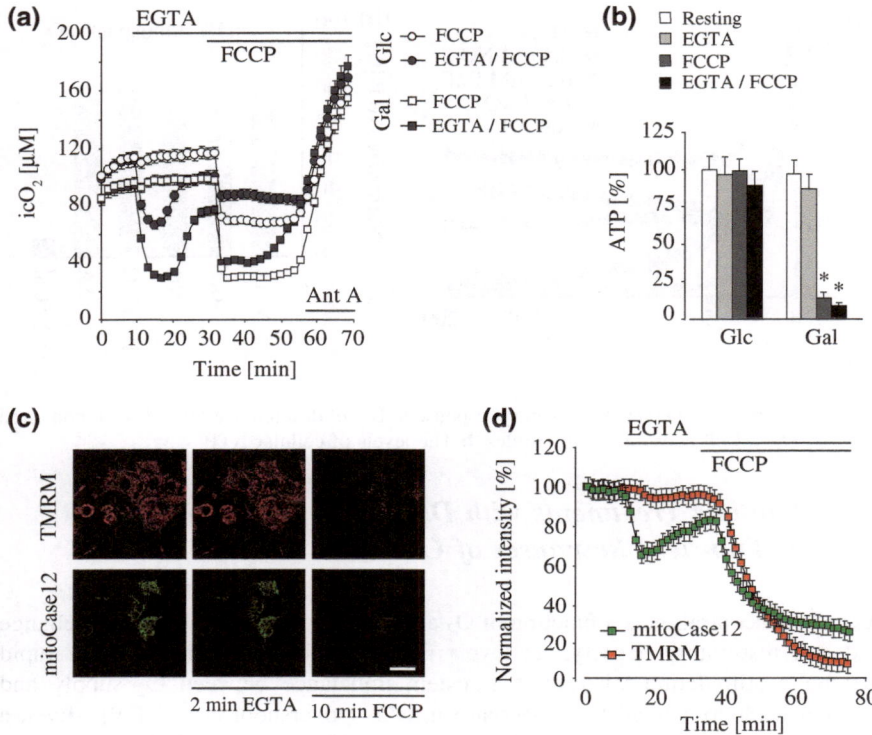

Fig. 2.21 Effect of eCa^{2+} chelation on the respiratory response of PC12 cells to FCCP. **a** Real-time profiles of icO_2. Replacement of glucose with galactose activates basal respiration and respiratory responses of the cells. Chelation of eCa^{2+} with 5 mM EGTA transiently increases cellular respiration. In eCa^{2+}-free conditions the response to mitochondrial uncoupling is reduced. Ant A inhibits respiration in all the samples and re-oxygenate the cells. **b** ATP levels measured at the peak of the response to FCCP (50 min) show a dramatic decrease in the absence of glycolytic ATP flux upon uncoupling. **c** Confocal images and **d** time profiles of TMRM and mitoCase12 fluorescence demonstrate the changes in $\Delta\Psi m$ and $mitoCa^{2+}$ upon EGTA and FCCP treatment in glucose(+) medium. Asterisks indicate significant difference. Scale bar represents 20 µm

cells treated with EGTA, the increase in respiration in response to FCCP was transient and followed by rapid reoxygenation of the cells reflecting their loss of respiratory activity. FCCP treatment caused a large drop in cellular ATP in galactose(+) medium, but not in glucose(+) (Fig. 2.21b). Interestingly, in the presence of eCa^{2+} the response to FCCP was more sustained suggesting that ATP levels are not critical for cell respiration. In the cells incubated on glucose and producing ATP via both glycolysis and OxPhos, the FCCP-specific increase in respiration was more sustainable, regardless of eCa^{2+}. Parallel confocal microscopy analysis revealed that depletion of eCa^{2+} caused a moderate decrease in $mitoCa^{2+}$ and only minor changes in $\Delta\Psi m$, whereas treatment with FCCP caused massive release of $mitoCa^{2+}$ and complete dissipation of $\Delta\Psi m$ (Fig. 2.21c,d). Pre-treatment with Ant A inhibited respiratory responses in all the samples.

2.4 Instrument Requirements and Selection Criteria

A wide variety of plate reader types are currently used in conjunction with extracellular probes for the analysis of cellular oxygen consumption [1, 5, 11, 20, 26, 31, 59, 62]. Such measurements are typically performed in prompt fluorescence mode or in TR-F mode. The advantage of the latter is that interference from autofluorescent components and optical properties of the test sample can be minimised significantly thereby providing superior signal to blank values and more stable readings. However, time resolution of such instruments is typically limited by the Xe-flashlamp used as excitation light source whereby the 'tail' is often still detectable at a delay of 20 µs while the emission lifetime of the probes themselves sets an upper limit on the delay time that can be used. A 30 µs delay is, therefore, recommended as an optimum value [11]. Another relevant consideration for TR-F readers is that detection is generally PMT based and which can be insensitive above 650 nm, although this is only problematic where probe signal is limited. The efficiency of plate heating can also be important, particularly when an entire microplate is being measured. Some instruments show a temperature differential across the plate whereby the centre of the plate is at a slightly lower temperature to the edge. Since both probe response and cell respiration are temperature sensitive, such temperature disparity can result in an 'edge effect'. This effect becomes less pronounced as one moves from 37 °C to 30 °C. Onboard data processing software also varies significantly across models. On some models, all data analyses and subsequent reductions can be done automatically on the plate reader software while in other extreme cases all data have to be exported to discrete package such as MS Excel for subsequent manual processing.

While an intensity-based measurement approach can be applied to certain extracellular oxygen measurements, the investigation of intracellular oxygen, extracellular pH or more quantitative extracellular oxygen consumption analysis, requires the RLD approach to lifetime determination outlined in Chap. 1. For this measurement mode, the choice of instrument is critical as both the lamp used and the S:B values achievable determine measurement success. In this regard, we have found Victor (PerkinElmer, Finland] and FLUOstar or POLARstar Omega^{TM} (BMG Labtech, Germany) to provide best performance with other, often higher end readers not showing the requisite performance. The addition of an Atmospheric Control Unit (ACU) to the FLUOstar Omega^{TM} allowing modulation of O_2 and CO_2 in the measurement chamber adds additional functionality which is particularly relevant to icO_2 investigations. These instruments work well with PtCP- and PtPFPP-based probes including MitoXpress$^®$, nanoparticles [10] and peptide conjugates [13] and are now used in the high-throughput assessment of O_2 in cell populations and other sample types. It also worth noting that, while this measurement approach is applicable to Pt-Porphyrin-based probes, it is far less successful in the measurements with Ru(II) complexes due to their short lifetime. Longwave O_2 probes emitting in the very-near infrared region can also be problematic for standard readers which are rather insensitive at >700 nm. There are

Table 2.2 Pt-porphyrin O_2 probe signals from samples with mammalian cells in standard 96-well plates measured across different instruments using recommended probe concentrations and settings optimised for kinetic analysis. These are indicative values as intensity signals can vary from instrument to instrument

Instrument	Probe	Intensity signals 1, 2[a]	Blanks 1, 2[a]	S:B	Measured lifetime, μs
Omega (BMG)	MitoXpress® (EC)	50,000/14,000	5,000/400	10/35	23.1 ± 0.5
	NanO2 (IC)	55,000/15,500	2,500/400	22/39	30.2 ± 0.5
Victor[4] (PerkinElmer)	MitoXpress® (EC)	150,000/26,800	700/265	210/120	23.2 ± 0.2
	NanO2 (IC)	150,000/40,100	700/265	210/145	30.3 ± 0.2
Genius Pro (Tecan)[b]	MitoXpress® (EC)	600	25	24	n/a
Spectramax[c] (Molecular Devices)	MitoXpress® (EC)	65	15	5	n/a

[a] Under optimised instrument settings and RLD delay times 30 and 70 μs, at 37 °C in air-saturated solution and optimised probe concentrations
[b] Instrument only supports TR-F, but not RLD mode
[c] can only measure Pt-porphyrins in prompt fluorescence mode
n/a—not applicable

also instances where instruments have been custom designed specifically for such applications and Pt-porphyrin labels [63].

Typical performance characteristics of several commercial fluorescent readers are summarised in Table 2.2. One can see that intensity signals and S:B ratio vary significantly from one instrument to another. Using non-optimal instruments, measurement and RLD settings may lead to a significant loss of performance of phosphorescence lifetime and O_2 concentration determination (inaccurate values, large S.D.). When analysing large number of samples on one plate, scanning speed of the instrument should be considered. On the first three instruments, one scan of a 96-well plate normally takes about 2 min. To increase measurement frequency (sampling time for each well), one can reduce the number of samples on the plate or change measurement settings (some loss of sensitivity may occur). Overall, for O_2 measurement with Pt-porphyrin probes strong preference is given to high sensitivity TR-F readers.

2.5   Conclusions

Overall, optical O_2 sensing using fluorescent and TR-F plate-reader detection represents a powerful and versatile technique for assessment of respiratory responses in cell populations. It also provides an experimental platform for the development and characterisation of new extracellular and intracellular O_2 probes, and facilitates the analysis of both O_2 concentrations and localised oxygen gradients. The utility of the technology is underscored by the spectrum of applications and experimental data described above demonstrating that optical O_2 sensing measurements performed on standard fluorescent/TR-F plate readers provide valuable information on the OCR, metabolic status, cell oxygenation and cellular responses to metabolic stimulation. Populations of cells in small samples placed in microtiter plates or other substrates can be measured in a simple, fast and high-throughput manner providing quantitative and statistically sound data. The choice of different measurement formats and detection modalities (described in Chap. 1) and the high degree of flexibility of such systems makes them well-suited for the analysis of cells systems and diverse biological and animal models under various external O_2 levels, metabolic stimulation or stress. Measurement can be performed in standard platforms such as microtiter plates or in customised systems such as capillary cuvettes, cell-based biochips and microfluidic systems. Experimental set-ups are relatively straightforward and do not require highly specialised instru-mentation, and are therefore broadly applicable across biological science even by the researchers without special skills or previous experience with optical O_2 sensing. A large number of academic and industrial labs have already adopted the technology for routine use in their research and screening.

In all these applications, modern TR-F readers usually provide high sensitivity in detecting the probes based on Pt-porphyrin dyes. However, not all instruments and probes can be used with the same degree of success. Instrument and probe must be mutually compatible whereby appropriate, spectral characteristics, time resolution, sensitivity, temperature control and measurement consistency must be achieved. Additional factors such as data processing software can also be important considerations when designing the experimental set-up.

Phosphorescent signals obtainable with the new generation icO_2 probes such as NanO2 can be as high as the signals with the extracellular probes. These probes give the user a considerable signal window to perform reliable and accurate lifetime-based O_2 and OCR measurements which can then be related to cell and sample parameters. Also these probes do not cause any significant cell damage or impact on cellular function. In some applications, even simple intensity-based fluorescent readers can provide very satisfactory analytical performance in mea-suring relative OCRs and relative oxygenation and changes in cell/sample respi-ration, however, lifetime-based O_2 sensing remains a preferred option.

Quantitative data generated in this manner with adherent cell cultures, icO_2 probes can be allied to a suite of additional functional bioassays providing a data set of high physiological relevance and giving more complete picture of cellular metabolic responses. The open microplate format used in icO_2 measurements allows multiple treatments to be applied in one experiment while tracing changes in icO_2 and cell respiration in real time. This is very useful for various mechanistic studies, when the interplay of different drugs, conditions and processes within the cell can be analysed. Moreover, phosphorescent O_2 sensing probes have high potential for multiplexing, and O_2 analyses can be coupled with parallel mea-surement of other cellular parameters, including ATP and NAD(P)H, cytosolic and mitochondrial Ca^{2+} and pH, the $\Delta\Psi m$ and redox state. This can again provide a more in-depth and detailed insight in cell metabolism and bioenergetics.

The two main approaches to O_2 measurement and corresponding set-ups complement each other. Thus, measurements with extracellular probes are usually conducted under oil overlay (i.e. partially sealed samples), in a kinetic assay which lasts 15–90 min producing one OCR value per sample. Effectors and treatments have to be applied prior to the assay, since additions during the measurement are inconvenient due to the oil seal. In contrast, O$_2$ measurements with intracellular probes are carried out in an open system (liquid sample is exposed to gaseous atmosphere), when repeatable additions of multiple effectors is applicable.

The format used for icO$_2$ measurement (see Sect. 2.3.1 and Fig. 2.13) resembles the tissue and blood vessel system of living organisms. Therefore, the processes observed in such assays (partial deoxygenation, dysbalance between O$_2$ utilization and supply) may also occur in vivo in a number of common (patho)physiological conditions such as ischemia and reperfusion, stroke, neuronal excitotoxicity, i.e. when certain areas remote from the source of O$_2$ undergo metabolic stress or activation, and also in cancer and embryo development. O$_2$ diffusion in tissue is thought to be similar to that of water, growth media and extracellular fluid [8, 64]. Characterisation of 3D O$_2$ gradients is, however, a more complicated task and requires specialised techniques, such as O$_2$ imaging microscopy described in Chap. 3. For larger specimens such as tissue sections and organs, this is even more challenging since optical techniques have limited penetration into tissue. These O$_2$ sensing techniques, therefore, have the potential to provide new insights into cellular function, mechanism of drug actions and even the examination of therapeutic targets and disease progression, particularly where mitochondria, metabolism or oxygen availability are central players.

Acknowledgments This work was supported by the Science Foundation Ireland, grant 07/IN.1/ B1804, and the EU FP7 project Chebana MC-IAPP-2009-230641.

References

1. Hynes J, Floyd S, Soini AE, O'Connor R, Papkovsky DB (2003) Fluorescence-based cell viability screening assays using water-soluble oxygen probes. J Biomol Screen 8:264–272
2. O'Riordan TC, Buckley D, Ogurtsov V, O'Connor R, Papkovsky DB (2000) A cell viability assay based on monitoring respiration by optical oxygen sensing. Anal Biochem 278:221–227
3. Schouest K, Zitova A, Spillane C, Papkovsky D (2009) Toxicological assessment of chemicals using Caenorhabditis elegans and optical oxygen respirometry. Environ Toxicol Chem 28:791–799
4. Zitova A, O'Mahony FC, Cross M, Davenport J, Papkovsky DB (2009) Toxicological profiling of chemical and environmental samples using panels of test organisms and optical oxygen respirometry. Environ Toxicol 24:116–127
5. Alderman J, Hynes J, Floyd SM, Krüger J, O'Connor R, Papkovsky DB (2004) A low-volume platform for cell-respirometric screening based on quenched-luminescence oxygen sensing. Biosens Bioelectron 19:1529–1535
6. Zitova A, Hynes J, Kollar J, Borisov SM, Klimant I, Papkovsky DB (2010) Analysis of activity and inhibition of oxygen-dependent enzymes by optical respirometry on the LightCycler system. Anal Biochem 397:144–151
7. O'Donovan C, Twomey E, Alderman J, Moore T, Papkovsky D (2006) Development of a respirometric biochip for embryo assessment. Lab Chip 6:1438–1444

8. Wilson DF (2008) Quantifying the role of oxygen pressure in tissue function. Am J Physiol Heart Circ Physiol 294:H11–H13

9. O'Riordan TC, Fitzgerald K, Ponomarev GV, Mackrill J, Hynes J, Taylor C, Papkovsky DB (2007) Sensing intracellular oxygen using near-infrared phosphorescent probes and live-cell fluorescence imaging. Am J Physiol Regul Integr Comp Physiol 292:R1613–1620

10. Fercher A, Borisov SM, Zhdanov AV, Klimant I, Papkovsky DB (2011) Intracellular O2 Sensing Probe Based on Cell-penetrating Phosphorescent Nanoparticles. ACS Nano 5:5499–5508

11. O'Riordan TC, Zhdanov AV, Ponomarev GV, Papkovsky DB (2007) Analysis of intracellular oxygen and metabolic responses of mammalian cells by time-resolved fluorometry. Anal Chem 79:9414–9419

12. Dmitriev RI, Papkovsky DB (2012) Optical probes and techniques for O(2) measurement in live cells and tissue. Cell Mol Life Sci [Epub ahead of print]

13. Dmitriev RI, Ropiak HM, Yashunsky DV, Ponomarev GV, Zhdanov AV, Papkovsky DB (2010) Bactenecin 7 peptide fragment as a tool for intracellular delivery of a phosphorescent oxygen sensor. FEBS J 277:4651–4661

14. Dmitriev RI, Zhdanov AV, Ponomarev GV, Yashunski DV, Papkovsky DB (2010) Intracellular oxygen-sensitive phosphorescent probes based on cell-penetrating peptides. Anal Biochem 398:24–33

15. Zhdanov AV, Dmitriev RI, Papkovsky DB (2011) Bafilomycin A1 activates respiration of neuronal cells via uncoupling associated with flickering depolarization of mitochondria. Cell Mol Life Sci 68:903–917

16. Zhdanov AV, Ward MW, Prehn JH, Papkovsky DB (2008) Dynamics of intracellular oxygen in PC12 Cells upon stimulation of neurotransmission. J Biol Chem 283:5650–5661

17. Zhdanov AV, Ward MW, Taylor CT, Souslova EA, Chudakov DM, Prehn JH, Papkovsky DB (2010) Extracellular calcium depletion transiently elevates oxygen consumption in neurosecretory PC12 cells through activation of mitochondrial Na(+)/Ca(2 +) exchange. Biochim Biophys Acta 1797:1627–1637

18. Hynes J, O'Riordan TC, Zhdanov AV, Uray G, Will Y, Papkovsky DB (2009) In vitro analysis of cell metabolism using a long-decay pH-sensitive lanthanide probe and extracellular acidification assay. Anal Biochem 390:21–28

19. Hynes J, Marroquin LD, Ogurtsov VI, Christiansen KN, Stevens GJ, Papkovsky DB, Will Y (2006) Investigation of drug-induced mitochondrial toxicity using fluorescence-based oxygen-sensitive probes. Toxicol Sci 92:186–200

20. O'Mahony F, Green RA, Baylis C, Fernandes R, Papkovsky DB (2009) Analysis of total aerobic viable counts in samples of raw meat using fluorescence-based probe and oxygen consumption assay. Food Control 20:129–135

21. Hynes J, Hill R, Papkovsky DB (2006) The use of a fluorescence-based oxygen uptake assay in the analysis of cytotoxicity. Toxicol in Vitro 20:785–792

22. Will Y, Hynes J, Ogurtsov VI, Papkovsky DB (2007) Analysis of mitochondrial function using phosphorescent oxygen-sensitive probes. Nat. Protocols 1:2563–2572

23. Yao J, Irwin R, Chen S, Hamilton R, Cadenas E, Brinton RD (2011) Ovarian hormone loss induces bioenergetic deficits and mitochondrial beta-amyloid. Neurobiol Aging 6(7):e21788

24. Hu L-F, Lu M, Tiong CX, Dawe GS, Hu G, Bian J-S (2010) Neuroprotective effects of hydrogen sulphide on Parkinson's disease rat models. Aging Cell 9:135–146

25. Chan DC (2006) Mitochondria: dynamic organelles in disease aging, and development. Cell 125:1241–1252

26. Marroquin LD, Hynes J, Dykens JA, Jamieson JD, Will Y (2007) Circumventing the Crabtree effect: replacing media glucose with galactose increases susceptibility of HepG2 cells to mitochondrial toxicants. Toxicol Sci 97:539–547

27. Kettenhofen R, Bohlen H (2008) Preclinical assessment of cardiac toxicity. Drug Discov Today 13:702–707

28. Choubey V, Safiulina D, Vaarmann A, Cagalinec M, Wareski P, Kuum M, Zharkovsky A, Kaasik A (2011) Mutant A53T α-synuclein induces neuronal death by increasing mitochondrial autophagy. J Biol Chem

29. O'Hagan KA, Cocchiglia S, Zhdanov AV, Tambuwala MM, Cummins EP, Monfared M, Agbor TA, Garvey JF, Papkovsky DB, Taylor CT, Allan BB (2009) PGC-1alpha is coupled to HIF-1alpha-dependent gene expression by increasing mitochondrial oxygen consumption in skeletal muscle cells. Proc Natl Acad Sci U S A 106:2188–2193

30. Favre C, Zhdanov A, Leahy M, Papkovsky D, O'Connor R (2010) Mitochondrial pyrimidine nucleotide carrier (PNC1) regulates mitochondrial biogenesis and the invasive phenotype of cancer cells. Oncogene 29:3964–3976

31. Hynes J, Swiss RL, Will Y (2012) High-throughput analysis of mitochondrial oxygen consumption. Methods Mol Biol 810:59–72

32. Sung HJ, Ma W, Wang PY, Hynes J, O'Riordan TC, Combs CA, McCoy JP Jr, Bunz F, Kang JG, Hwang PM (2010) Mitochondrial respiration protects against oxygen-associated DNA damage. Nat Commun 1:5

33. Jonckheere AI, Huigsloot M, Janssen AJ, Kappen AJ, Smeitink JA, Rodenburg RJ (2010) High-throughput assay to measure oxygen consumption in digitonin-permeabilized cells of patients with mitochondrial disorders. Clin Chem 56:424–431

34. Kelm JM, Lorber V, Snedeker JG, Schmidt D, Broggini-Tenzer A, Weisstanner M, Odermatt B, Mol A, Zünd G, Hoerstrup SP (2010) A novel concept for scaffold-free vessel tissue engineering: self-assembly of micro tissue building blocks. J Biotechnol 148:46–55

35. Amacher DE (2005) Drug-associated mitochondrial toxicity and its detection. Curr Med Chem 12:1829–1839

36. Lin MT, Beal MF (2006) Mitochondrial dysfunction and oxidative stress in neurodegenerative diseases. Nature 443:787–795

37. Muller WE, Eckert A, Kurz C, Eckert GP, Leuner K (2011) Mitochondrial dysfunction: common final pathway in brain aging and Alzheimer's disease–therapeutic aspects. Mol Neurobiol 41:159–171

38. Bartlett K, Eaton S (2004) Mitochondrial beta-oxidation. Eur J Biochem 271:462–469

39. Tennant DA, Duran RV, Gottlieb E (2010) Targeting metabolic transformation for cancer therapy. Nat Rev Cancer 10:267–277

40. Wise DR, DeBerardinis RJ, Mancuso A, Sayed N, Zhang XY, Pfeiffer HK, Nissim I, Daikhin E, Yudkoff M, McMahon SB, Thompson CB (2008) Myc regulates a transcriptional program that stimulates mitochondrial glutaminolysis and leads to glutamine addiction. Proc Natl Acad Sci U S A 105:18782–18787

41. O'Flaherty L, Adam J, Heather LC, Zhdanov AV, Chung YL, Miranda MX, Croft J, Olpin S, Clarke K, Pugh CW, Griffiths J, Papkovsky D, Ashrafian H, Ratcliffe PJ, Pollard PJ (2010) Dysregulation of hypoxia pathways in fumarate hydratase-deficient cells is independent of defective mitochondrial metabolism. Hum Mol Genet 19:3844–3851

42. Zhdanov AV, Favre C, O'Flaherty L, Adam J, O'Connor R, Pollard PJ, Papkovsky DB (2011) Comparative bioenergetic assessment of transformed cells using a cell energy budget platform. Integr Biol (Camb) 3:1135–1142

43. Ferrick DA, Neilson A, Beeson C (2008) Advances in measuring cellular bioenergetics using extracellular flux. Drug Discov Today 13:268–274

44. Hynes J, Natoli E, Jr, Will Y (2009) Fluorescent pH and oxygen probes of the assessment of mitochondrial toxicity in isolated mitochondria and whole cells. Curr Protoc Toxicol Chap. 2, Unit 2 16

45. Ashrafian H, O'Flaherty L, Adam J, Steeples V, Chung YL, East P, Vanharanta S, Lehtonen H, Nye E, Hatipoglu E, Miranda M, Howarth K, Shukla D, Troy H, Griffiths J, Spencer-Dene B, Yusuf M, Volpi E, Maxwell PH, Stamp G, Poulsom R, Pugh CW, Costa B, Bardella C, Di Renzo MF, Kotlikoff MI, Launonen V, Aaltonen L, El-Bahrawy M, Tomlinson I, Pollard PJ (2010) Expression profiling in progressive stages of fumarate-hydratase deficiency: the contribution of metabolic changes to tumorigenesis. Cancer Res 70:9153–9165

46. Wamelink MM, Struys EA, Jakobs C (2008) The biochemistry, metabolism and inherited defects of the pentose phosphate pathway: a review. J Inherit Metab Dis 31:703–717
47. Launonen V, Vierimaa O, Kiuru M, Isola J, Roth S, Pukkala E, Sistonen P, Herva R, Aaltonen LA (2001) Inherited susceptibility to uterine leiomyomas and renal cell cancer. Proc Natl Acad Sci U S A 98:3387–3392
48. Knox C, Sass E, Neupert W, Pines O (1998) Import into mitochondria, folding and retrograde movement of fumarase in yeast. J Biol Chem 273:25587–25593
49. Tomlinson IP, Alam NA, Rowan AJ, Barclay E, Jaeger EE, Kelsell D, Leigh I, Gorman P, Lamlum H, Rahman S, Roylance RR, Olpin S, Bevan S, Barker K, Hearle N, Houlston RS, Kiuru M, Lehtonen R, Karhu A, Vilkki S, Laiho P, Eklund C, Vierimaa O, Aittomaki K, Hietala M, Sistonen P, Paetau A, Salovaara R, Herva R, Launonen V, Aaltonen LA (2002) Germline mutations in FH predispose to dominantly inherited uterine fibroids, skin leiomyomata and papillary renal cell cancer. Nat Genet 30:406–410
50. http://www.ibidi.com/products/p_disposables.html
51. Zitova A, Cross M, Hernan R, Davenport J, Papkovsky DB (2009) Respirometric acute toxicity screening assay using Daphnia magna. Chem Ecol 25:217–227
52. Zitova A, O'Mahony FC, Cross M, Davenport J, Papkovsky DB (2009) Toxicological profiling of chemical and environmental samples using panels of test organisms and optical oxygen respirometry. Environ Toxicol 24:116–127
53. Schouest K, Zitova A, Spillane C, Papkovsky DB (2009) Toxicological assessment of chemicals using caenorhabditis elegant and optical oxygen respirometry. Environ Toxicol Chem 28:791–799
54. Zitova A, O'Mahony FC, Kurochkin IN, Papkovsky DB (2010) A simple screening assay for cholinesterase activity and inhibition based on optical oxygen detection. Anal Lett 43:1746–1755
55. Jezek P, Plecita-Hlavata L, Smolkova K, Rossignol R (2010) Distinctions and similarities of cell bioenergetics and the role of mitochondria in hypoxia, cancer, and embryonic development. Int J Biochem Cell Biol 42:604–622
56. Linsenmeier RA (1986) Effects of light and darkness on oxygen distribution and consumption in the cat retina. J Gen Physiol 88:521–542
57. Metzen E, Wolff M, Fandrey J, Jelkmann W (1995) Pericellular PO2 and O2 consumption in monolayer cell cultures. Respir Physiol 100:101–106
58. Rumsey WL, Schlosser C, Nuutinen EM, Robiolio M, Wilson DF (1990) Cellular energetics and the oxygen dependence of respiration in cardiac myocytes isolated from adult rat. J Biol Chem 265:15392–15402
59. Zhdanov AV, Ogurtsov VI, Taylor CT, Papkovsky DB (2010) Monitoring of cell oxygenation and responses to metabolic stimulation by intracellular oxygen sensing technique. Integr Biol 2:443–451
60. Bowman EJ, Siebers A, Altendorf K (1988) Bafilomycins: a class of inhibitors of membrane ATPases from microorganisms, animal cells, and plant cells. Proc Natl Acad Sci U S A 85:7972–7976
61. Teplova VV, Tonshin AA, Grigoriev PA, Saris NE, Salkinoja-Salonen MS (2007) Bafilomycin A1 is a potassium ionophore that impairs mitochondrial functions. J Bioenerg Biomembr 39:321–329
62. Schoonen WG, Stevenson JC, Westerink WM, Horbach GJ (2012) Cytotoxic effects of 109 reference compounds on rat H4IIE and human HepG2 hepatocytes. III: Mechanistic assays on oxygen consumption with MitoXpress and NAD(P)H production with Alamar Blue. Toxicol in Vitro 26:511–525
63. O'Riordan TC, Soini AE, Soini JT, Papkovsky DB (2002) Performance evaluation of the phosphorescent porphyrin label: solid-phase immunoassay of alpha-fetoprotein. Anal Chem 74:5845–5850
64. Kapellos GE, Alexiou TS, Payatakes AC (2007) A multiscale theoretical model for diffusive mass transfer in cellular biological media. Math Biosci 210:177–237

Chapter 3
O$_2$ Imaging in Biological Specimens

Andreas Fercher, Alexander V. Zhdanov and Dmitri B. Papkovsky

Abstract This chapter describes the fundamentals of an O$_2$ imaging technique based on the quenched-phosphorescence detection of Pt-porphyrin probes. The wide-field, confocal and multi-photon microscopy and methodological aspects of quenched phosphorescence icO$_2$ imaging techniques, theoretical and practical considerations, are briefly described and critically assessed. This is followed by a comprehensive set of practical examples in which the imaging of various biological models, including conventional cell cultures, spheroids (neurospheres), larger organisms such as *C. elegans* worms and microfluidic devices were analysed. Critical factors that determine the performance of such imaging experiments are also identified and discussed providing a broad prospective on the possible applications of these techniques, particularly in the studies of cell and tissue physiology, the role of O$_2$ in metabolism, hypoxia and other areas of biomedical research.

Keywords Live cell imaging, oxygen imaging · Phosphorescence quenching · Microsecond FLIM · Confocal and multi-photon microscopy · Cellular oxygen · Intracellular probes

3.1 Introduction

Fluorescence live cell imaging is the core bioanalytical technique used routinely in many areas of life and biomedical sciences [1]. Fluorescence microscopy generally meets the demand of sensitivity, selectivity and spatial resolution, and is applicable to a broad range of biological models, including individual mammalian cells, cell

D. B. Papkovsky et al., *Phosphorescent Oxygen-Sensitive Probes*,
SpringerBriefs in Biochemistry and Molecular Biology,
DOI: 10.1007/978-3-0348-0525-4_3, © The Author(s) 2012

populations, heterogeneous tissues, organs and even whole animals. It allows studies of morphology, membrane and organelle structures, functional characteristics of the cells, metabolic status and processes such as ion fluxes, excitability and signalling [2–6]. Using time-lapse imaging, dynamic changes in these parameters can be monitored with a high level of detail and spatial resolution. Imaging techniques are constantly evolving, with many new probes, cellular assays and measurements modalities developed in recent years including a group of in vivo and super-resolution imaging methods [7, 8]. An inherent constraint associated with such imaging techniques is that of limited penetration depth into live tissue, however, new long wave probes, detection modalities (multi-photon excitation) and technological developments constantly extend this and broaden the analytical capabilities of optical bioimaging.

Despite the important roles of O$_2$ in biological systems, currently its analysis in cells and tissues is not performed routinely. This is due to a number of factors which include: (i) special features of O$_2$ detection methods in comparison, to the other analytes (see Chap. 1); (ii) the limited number of O$_2$ probes currently available for imaging applications and/or their non-optimal performance; (iii) a frequent need to modify and custom-tune standard imaging hardware and software for available O$_2$ probes including adaptations for microsecond FLIM readout to enable generation of images in O$_2$ concentration scale [9]; (iv) the lack of optimised and validated procedures for O$_2$ imaging in biological samples of interest. On the other hand, adaptation of existing fluorescent imaging platforms for O$_2$ imaging is relatively straightforward. A number of specialised systems and applications have been reported in recent years, producing exciting experimental data and revealing important biological findings demonstrating the opportunities for O$_2$ imaging in both basic and translational biomedical research [1, 10].

As previously stated, intensity-based microscopy has limitations, due to broad variations in luminophore concentration and distribution within and between samples, background signals (scattering autofluorescence), luminophore photobleaching, instability of the light source and detector sensitivity. These factors influence measured signals and O$_2$ calibration and complicate quantitative measurements whereby detailed controls are required for each experiment. Ratiometric intensity imaging at two different wavelengths can compensate for many of these variables and provide a more stable calibration, in which O$_2$ concentration is calculated from the ratio of the O$_2$-sensitive and reference luminescent signals in two spectral regions [13, 14]. However, this ratio can also be influenced by the sample and measurement system parameters such as differential photobleaching, scattering and autofluorescence in the two spectral channels, detector noise and blank signals. A lifetime imaging modality (FLIM) overcomes these limitations since phosphorescence lifetime (τ) is an intrinsic parameter of the probe which is rather insensitive to instrumentation factors, probe concentrations, photobleaching and optical alignment [11], but sensitive to changes in O$_2$ concentration in probe microenvironment.

Time-domain FLIM on wide-field microscopes equipped with gated CCD camera and pulsed LED excitation is simple. It is well-suited for measuring the long-decay emission of Pt-porphyrin-based probes and provides 2D lifetime and O_2 images with good sensitivity and temporal resolution. Generation of 3D images is also possible, but this requires additional hardware and complex data processing software. Laser-scanning confocal FLIM together with time-correlated single photon counting (TCSPC) adds another dimension and allows detailed O_2 mapping of 3D specimens. For quantitative O_2 detection the system must be calibrated with O_2 standards under defined measurement conditions with consideration for parameters such as temperature and sample type.

The initial appraisal of O_2 imaging has been with extracellular (cell-impermeable) probes injected into the vasculature or animal tissue which have to stay in the circulation during the measurement [10, 12, 13, 14]. The new generation of intracellular O_2 sensing probes, which provide passive self-loading of live mammalian cells and tissues and produce high and stable phosphorescent signals, have extended the capabilities of O_2 imaging. Such probes enable once-off loading of samples through addition to the medium followed by a short staining period after which time cells remain stained for long periods of time and can be analysed without the need to keep the probe at high concentration in the bathing solution [12, 15]. These probes are well-suited to in vitro and ex vivo studies with various biological models including individual cells (prokaryotes and eukaryotes), layers or adherent cells, microbial biofilms, spheroids, mixed cell cultures, engineered tissue, perfused animal organs and whole organisms and have a potential for use in live higher organisms. Intracellular O_2 probes therefore can inform on cell and tissue oxygenation as well as changes in respiration and cellular function under different physiological conditions including hypoxia, metabolic stimulation and various disease states.

In this chapter we describe the main principles of live cell imaging, mainly with Pt-porphyrin-based icO_2 probes and nanosensors, and the practical uses of these probes and imaging techniques in life science research. A number of examples with different biological models are presented, for which oxygenation conditions, O_2 distribution and kinetics are assessed. These experiments are conducted using the wide-field CCD camera-based FLIM, laser-scanning confocal TCSPC based FLIM and two-photon ratiometric intensity measurements to image and trace O_2 concentrations.

3.2 Wide-Field Microsecond FLIM

In wide-field microscopy commonly used with small biological samples, the whole field of view is illuminated with a light source while fluorescence is collected from the specimen by the optics and visualised on the ocular or an appropriate imaging detector such as charged couple device (CCD) camera. The emitted light is separated from the excitation by a set of optical filters and dichroic mirror. Mercury arc lamps, lasers or light emitting diodes (LEDs) are used to excite reporter molecules in the specimen with emission light being a conglomerate from all the space in the focal plane and from excited regions above and below the focal plane.

Therefore, the contrast of wide-field microscopy is lower than for laser-scanning confocal microscopy. Despite this it provides high quality images with good sub-cellular detalisation as well as sub-micrometer spatial and millisecond temporal resolution.

FLIM can be performed with time-gated CCD cameras coupled with modulated excitation in the time (TD) or frequency domains (FD). FLIM measurements with conventional fluorophores having nanosecond lifetimes require high-speed light sources, detectors and electronics, and correction for background signals and light source function that could influence the calculated lifetime [16]. For Pt-porphyrin-based probes (lifetimes 10–100 μs) operational requirements for opto-electronic components are not as strict and the potential impact of such parameters on life-time determination is small. As a result, phosphorescence lifetime measurements with O$_2$ probes are technically more straightforward. Background signals, even if high, decay completely within the first 100 ns after the excitation pulse after which time the true lifetime of the O$_2$ probe can be reliably measured and accurately correlated with O$_2$ concentration [17].

Super-bright LEDs represent a simple solution for wide-field FLIM/PLIM systems [18]. They are affordable, can be modulated electronically with an LED driver to produce pulses or periodic modulation, are available for different spectral regions to excite all common luminophores, are reliable, stable, easy to handle and require no cooling. For optimal FLIM with Pt-porphyrins, excitation pulse width and frequency have to be in the range 1–10 μs and/or 1–10 kHz, respectively [16]. Mercury arc lamps commonly used in steady-state microscopy are not very suit-able for microsecond FLIM. Also for continuous and high-frequency lasers opto-acoustical modulators have to be used [19].

Gated CCD cameras are also common in wide-field microscopy and FLIM. Cameras with image intensifiers are used to detect weak luminescent signals, but they are more costly, delicate and often have reduced resolution and increased noise signals due to loss of photons during the conversion and generation of a cascade of electrons on the secondary emitter. FD-FLIM for luminophores having lifetimes of up to 1 ms can also be realised, using modulated LED excitation and detection of emission phase shift with an intensified CCD camera [8, 9]. FD-FLIM generally requires higher signal intensities and is more prone to interference by autofluorescence and scattering than TD-FLIM. Measurement at several modula-tion frequencies can be used to devise optical background and multi-exponential luminescence decays [10].

With these considerations in mind, O$_2$ imaging by wide-field microsecond FLIM is relatively easy to set up for the monitoring of oxygenation and transient local changes in O$_2$ concentration in 2D. Phosphorescent images can be taken over the whole field of view, image acquisition rates are faster than for laser-scanning microscopy (though significantly slower than for nanosecond FLIM) providing information-rich data. Commercial TD-FLIM systems suitable for O$_2$ imaging with Pt-porphyrin probes with good temporal dynamic range and general sim-plicity are available at affordable prices [18].

3.3 Confocal Microsecond FLIM and Multi-Photon O$_2$ Imaging

Fluorescent confocal laser scanning microscopy allows reconstruction of sharp 3D images of samples using the principle of double spatial filtering in which only focused excitation and emission light passes the pinhole and reaches the detector [20, 21]. The laser scans the specimen point-by-point in X–Y-coordinates and plane-by-plane in Z-coordinates thus producing 3D images of the intensity signal reflecting luminophore distribution within the sample.

Combined with photon counting detectors and TCSPC (time-correlated single photon counting) technique, confocal microscopy enables recording of low-level luminescent signals in FLIM mode with high spatial precision [9]. The distribution of photons in the sample can be recorded at count rates of up to 10 MHz per SPC module, thus allowing FLIM in a broad ns-ms range. Using high-frequency pulsed 405 nm diode laser operating synchronously with the pixel clock of the scanner, the phosphorescent signal is recorded by the TCSPC during the off-phase while a decay curve is constructed and lifetime value determined for every pixel of the image. Depending on the configuration and operation mode, TCSPC confocal systems can record single plane FLIM/PLIM images, sequences of images (in Z- or/and time axis) and multi-spectral images of Pt-porphyrin-based O$_2$-sensing probes in the μs time range and of other molecular probes in the ns time range. By applying an O$_2$ calibration, lifetime images can be converted into detailed patterns of O$_2$ distribution within a single cell or bigger samples such as spheroids or tissue sections.

The penetration depth of confocal imaging is not always sufficient, and for larger 3D objects multi-photon microscopy (MPM) is more preferred. In MPM the sample is excited with a focused near-infrared femtosecond laser producing high photon density so that the luminophore can simultaneously absorb two or more photons. Probe excitation therefore occurs only at the diffraction limited focal point and out-of-focus luminescence is eliminated [20]. MPM provides improved sensitivity and resolution with thick tissue sections and reduced scattering while longwave light is less damaging to live cells and can penetrate deeper into the sample. The approach can also be combined with TCSPC and FLIM to enable quantitative O$_2$ imaging [22, 23].

3.4 Sample Preparation for Quantitative O$_2$ Imaging

Performing O$_2$ imaging on live cells or organisms by the phosphorescence quenching technique requires consideration of a number of parameters. Sensitive biological samples must be handled appropriately and measured under well-defined and physiologically relevant conditions. An incubation chamber must be included in the imaging system to provide stable and accurate control of environmental parameters such as temperature, humidity, CO$_2$ and O$_2$ content.

Reliable control of these parameters is also critical for successful O_2 imaging, as variation will have a profoundly negative impact on the quality of measurements.

3.4.1 Sample Temperature

Respiration of biological specimens and probe response are both strongly dependent on temperature therefore respiring samples have to be analysed at physiological temperatures. Proper equilibration and reliable control of sample temperature must be provided prior to and during sample imaging. Microscopes dedicated for live cell imaging are usually equipped with heated sample stage, objectives and incubation chambers which provide circulation of pre-warmed, humidified gaseous atmosphere to quickly equilibrate and maintain a constant sample temperature. In addition, heating of both the sample dish and the objective is useful, especially for oil immersion objectives, as it prevents temperature related defocusing. Temperature stability is increased by using larger sample volumes or an additional reservoir with liquid on the imaging stage. The latter acts as a buffer and helps reduce sample evaporation and temperature fluctuations thereby increasing in-focus stability in long-term imaging sessions. In experiments requiring sample manipulation such as addition of effector, change of medium or gaseous atmosphere, associated fluctuations of temperature within the sample must be considered, and related effects on the measured parameters corrected through the incorporation of appropriate controls.

3.4.2 Properties of the Medium and pH

Biological samples usually require a defined medium composition. Thus, mammalian cells and tissues are normally cultured in a medium supplemented with bicarbonate and a humidified atmosphere containing 5 % CO_2. In long-term imaging experiments strict pH control is critical to avoid significant acidification of the medium, especially by highly glycolytic cancer cell lines. To prevent strong acidification in the absence of CO_2 control, growth medium should be replaced with an appropriate CO_2-free buffer system such as 10–25 mM HEPES. Probe phosphorescent signals (intensity and lifetime) can also be modulated by media components such as phenol red or evaporation-related alterations in ionic strength or viscosity. Such parameters can also have a significant influence on cell respiration and on the O_2 calibration of the probe being used. If corresponding information is not available from the literature, this needs to be checked experimentally.

3.4.3 Probe Photostability, Cyto- and Phototoxicity

Probe photostability is one of the main parameters in O_2 imaging, especially for kinetic and long-term experiments. Probe photobleaching occurs due to irreversible damage of the phosphorescent molecules upon excitation. It has to be considered very carefully in intensity-based imaging, and even in FLIM significant photobleaching can affect lifetime determination and gradually increase error in lifetime and O_2 calculation due to a decrease in S:N ratio. By-products of probe excitation, mainly singlet oxygen, can also cause collateral photodamage to the dye and cellular structures (lipids, proteins and nucleic acids). This usually occurs within a 50 nm radius from the photosensitiser, a distance determined by the lifetime and diffusion coefficient of singlet oxygen (4 µs in water and < 0.5 µs in tissue [24]. Photobleaching of the probe and sample photodamage can be reduced by minimising the intensity of excitation, exposure time, the number of frames and sampling frequency.

3.5 Loading of O_2 Probes

Sample preparation for O_2 imaging usually involves the loading of cells or biological sample with an icO_2 probe and the subsequent removal of excess unloaded probe by washing with fresh medium. The main probe types and loading approaches are described in previous chapters. Different probes require different loading conditions (doses, incubation times, special medium) while a suitable transfection strategy must also be chosen with optimisation of the use of transfection reagent, osmotic shock, microinjection or gene gun. Self-loading can also be achieved through the rational probe design including the use of cell-penetrating probes, and such approaches can produce efficient, convenient and reproducible loading. A number of such probes have been developed in recent years, some of which are produced commercially.

For example, NanO2 [25] (produced commercially by Luxcel Biosciences) and its analogue MM2 probe [26] provide efficient cell loading by simple addition to the medium at low concentrations (2–10 µg/ml) and incubation for 3–16 h at 37 °C in standard conditions (regular growth media, 5 % CO_2 atmosphere). Possessing high brightness and photostability in loaded cells these probes are well-suited for FLIM and even intensity imaging allowing for rapid repetitive signal acquisition with short exposure times [25]. Convenient spectral characteristics (excitation bands at 380–405 nm and 520–545 nm, and emission at 650 nm) make NanO2 compatible with standard live cell imaging systems, while MM2 probe has an additional FRET antennae allowing two-photon and ratiometric intensity imaging [26].

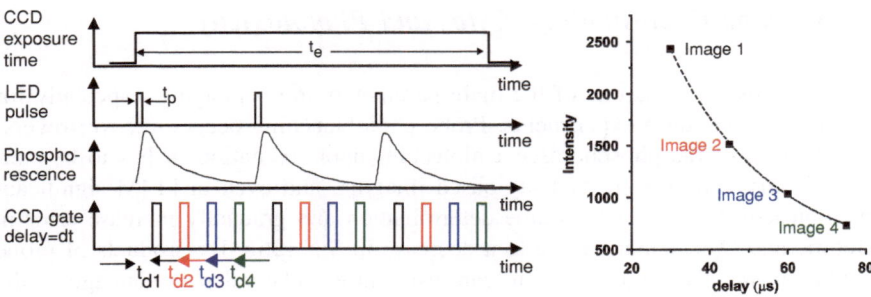

Fig. 3.1 Measurement scheme for the wide-field CCD camera-based TD-FLIM

3.6 Settings for O$_2$ Imaging on Camera-Based FLIM

The relatively simple wide-field TD-FLIM system set discussed here consists of a standard inverted fluorescence microscope Axiovert 200 (Carl Zeiss) equipped with a fast-gated CCD camera Imager Pro (PCO/LaVision Biotech) and an LED excitation module. The excitation module contains 3 super-bright LEDs (390 nm, 470 nm and 590 nm, selected manually by the operator) and an LED driver (LaVision), allowing the generation of excitation pulses of different length and frequency and synchronisation with the CCD camera. With this set of LEDs, microsecond FLIM of several common O$_2$ probes can be performed, including the PtCP, PtPFPP (excitable with 390 nm LED, emission band at 630–670 nm), PtCPK and PtBP (excitable with 390 nm or 590 nm LEDs, emission band at 740–800 nm) based probes [12]. Pd-porphyrins as well as conventional fluorophores such as DAPI, FITC, GFP, rhodamines can also be measured. The imaging system is equipped with an incubator system (Incubator S, Pecon, Germany) which allows precise control of sample temperature, humidity, CO$_2$ and O$_2$ content.

The principle of acquisition of phosphorescent signals in FLIM mode is shown schematically in Fig. 3.1. To ensure accurate generation of phosphorescence decay, fitting and lifetime determination, FLIM requires preliminary optimisation of measurement conditions and signal acquisition parameters. Optimally, FLIM should be used with a S:N > 3 within the field of view or regions of interests (ROI). This corresponds to phosphorescence intensity signals of at least 500–1,000 counts (for the shortest delay time in a series) and blanks of about 100–300. The CCD camera reaches saturation at 16,000 counts per pixel.

The following main parameters of the FLIM system determine the magnitude of signals in primary intensity images: (1) Exposure Time (t_e) which is defined by the overall data acquisition time per frame; (2) Pulse Length (t_p) defines the duration of excitation pulses generated periodically during the exposure time; (3) Repetition Time (t_r) defines the frequency of excitation pulses; (4) Gate Time (t_g) defines time window after the excitation pulse when phosphorescence is collected (constant for each delay time); (5) Delay Time (t_d) defined by the time of emission decay (μs range) and the number of frames or data points required for fitting the

Table 3.1 Typical settings used on the wide-field FLIM for O₂ imaging of adherent mammalian cells with Pt-porphyrin probes and 40x oil immersion objective, NA 1.3 (Carl Zeiss)

Parameter	Settings
Exposure time (t_e)	50–2,000 ms
Pulse length (t_p)	10 μs
Decay curve start time	30 μs
Decay curve end time	150 μs
Number of frames	5–11
Gate time (t_g)	10–30 μs
Repetition time (t_r)	170 μs
Binning	2 × 2

decay curve; (6) Binning allows grouping of several pixels, thus decreasing spatial resolution but increasing signal intensity and improving temporal resolution. These parameters further depend on the application requirements, the type, magnification and numerical aperture of the objective, sample type and probe used.

Table 3.1 shows a typical set of instrumental parameters used for O₂ imaging with the Pt-porphyrin probes on the wide-field FLIM system described above. The effects of these parameters have been assessed individually and in a combinatorial fashion to optimise them for the different probes and biological samples measured on this system. Under these settings, a set of phosphorescence intensity images (frames) are recorded at different delay times, **t_d**. Each image represents counts integrated over the defined **t_g** and **t_e** which are maintained constant. The resulting set of frames is processed using single or double-exponential decay fits ($I = I_o*\exp(-t_d/\tau) + I_b$ or $I = I_1*\exp(-t_d/\tau_1) + I_2*\exp(-t_d/\tau_2) + I_b$, respectively) to generate lifetime values for each pixel/bin of interest and then to construct a lifetime image. Arithmetic transformation of the lifetime image using the calibration function (determined in a separate experiment), allows reconstruction of a 2D image of O₂ distribution within the sample. Such sets of experiments can then be repeated to follow the changes in O₂ distribution over time (time lapse mode).

3.7 Acquisition of Images in Wide-Field FLIM Mode

The system requires initial warm-up and equilibration for the camera and incubator (temperature, CO_2, O_2) for about 30 min. After placing the sample on the imaging stage and temperature equilibration for ∼ 30 min, pre-focusing and selection of appropriate ROI is performed in transmission mode, using short exposure time of the CCD camera (∼ 5 ms) and excitation with halogen lamp. Focusing in fluorescence mode (1–2 frames) is then performed to set up the sample for time-lapse intensity or FLIM measurements and to inspect probe distribution and loading efficiency. During these preparations, exposure of sample to excitation light should be kept to a minimum to minimise probe photobleaching and phototoxicity effects. After this, if all the settings are deemed satisfactory (tuning can be performed, if required) and phosphorescence intensity signals are reliably measurable, FLIM images can be acquired. In addition, changes in probe signal can be monitored in time lapse intensity mode, which reflect relative changes in O$_2$ concentration within the sample and also probe photobleaching. If an automated time-lapse FLIM recording is not available, the snapshot time points for FLIM measurements are worked out, based on the predictions of the physiological responses (changes cellular function and oxygenation as a result of drug treatment), or the results of time lapse intensity measurements.

To measure O$_2$ distribution and biological responses in a statistically sound fashion, imaging experiments have to be repeated several times to ensure consistency and reproducibility. For gentle addition of drugs/effectors with minimal distortion of the sample and local O$_2$ gradients within glass bottom Petri minidishes, a dispensing micro-pump can be used. Normally effector stock is loaded with a Hamilton syringe into the tubing of the pump (10x concentration, 1/10 of sample volume), allowed to equilibrate temperature and gas composition and then pumped into the dish with sample sitting on the imaging stage inside the incubation chamber. Effector additions are usually done at a low pumping rate, after a stable baseline of icO$_2$ is achieved in the sample with resting cells ($\sim$ 20 min of monitoring or incubation). A microfluidic chip system overcomes the cumbersome effector addition by pumping the effector containing medium through microchambers with cells.

When data acquisition is completed, lifetime image for the whole field of view or particular ROI is generated as described above (see Sect. 3.6), using the ImSpector software (LaVision). For the fitting, signal threshold is usually set above average background signal and below average probe signal (normally $\sim$ 3–5 times the background), while offset signal in the decay is calculated separately for each pixel. Smoothening of the image by 3 $\times$ 3 or 5 $\times$ 5 pixel binning reduces the noise within the image, but leads to a more pixelated image. The lifetime image can be exported on a pixel-to-pixel basis and further analysed within the software by generating line profiles or histograms over selected areas. This process has to be done for each generated lifetime image or stack of images (for time lapse experiments). After this, by applying the O$_2$ calibration (see below), O$_2$ maps can be produced on a pixel-by-pixel basis, e.g. using MatLab software and exported as O$_2$ concentration images. Clearly, such multi-step processing of FLIM images can be streamlined by designing automated data processing software and tailoring it to each particular application, sample or measurement task.

3.8 Calibration

In order to convert measured lifetime (or intensity) images into O$_2$ concentration maps, calibration under the defined measurement conditions has to be carried out. Some probes retain stable calibrations with and without cells, while others show some specificity with respect to the specimen, and medium used. Measurement instrumentation can also bring some variability and systematic errors. If obtained from another source (e.g. probe manufacturer or different user), calibration verification is recommended for each new imaging system and/or biological specimen.

The calibration procedure is relatively straightforward and is similar for different biological specimens and conditions. Here we present one example for cultured mammalian cells, a 40x oil objective (NA 1.3) and physiological temperature. Similar procedure is described in [25] and also in Chap. 2. Briefly, cells are loaded with probe and then incubated with 10 µM Antimycin A for 30 min to inhibit their respiration and bring icO_2 to atmospheric pO_2 levels. The pO_2 on the sample stage is controlled by the incubation system with closed loop gas circulation and built-in O_2 sensor. For each standard O_2 level, cells are incubated for about 30 min to allow gas equilibration prior to the imaging. After this, the sample is focused in transmission and then fluorescence intensity mode by taking 1–2 images (3×3 pixels). FLIM measurements are then performed in triplicates for each O_2 standard. For each calibration point, FLIM measurements are conducted on three different fields of the sample, ensuring that they have similar probe signals. Since the majority of hypoxia chambers and incubator units cannot provide accurate O_2 levels in the atmosphere below 0.5–1.0 % O_2, the 0 % O_2 condition can be achieved by adding 100 µg/mL of glucose oxidase and 10–25 mM glucose to the medium. This causes fast, deep and sustained deoxygenation of the sample bringing lifetime values to anticipated levels. The temperature should be maintained constant, normally at 37 °C for mammalian cells.

Within lifetime images several ROI's are analysed to determine average lifetime values. These values are then assigned to the given O_2 standard concentration and used as data points for the calibration. An example of O_2 calibration for the NanO2 probe is shown in Fig. 3.2.

Experimental data points of O_2 calibration are then best fitted ($R^2 > 0.98$) with a single-exponential decay function, to determine the analytical relationship between measured τ (µs) and icO_2 concentration (µM). Thus, for the results presented in Fig. 3.2a, the following calibration function was obtained:

$$[icO2] = 2944.66_{*e} \left(-\frac{\tau}{11.03} \right) - 5.54 \qquad (3.1)$$

The Stern–Volmer plot for τ_0/τ showed good linearity in the physiological O_2 range with K_{sv} of 5.41 ± 0.20 µM^{-1}. It is also worth noting that intensity signals increase ~ 4 fold from 1,500 counts to 5,800 counts when going from 20 to 0 % of atmospheric O_2.

Besides adherent mammalian cells, O_2 calibrations have also been performed with SK-N-AS neurospheres, biofilms of *P. aeruginosa* loaded with NanO2 probe, concentrated NanO2 probe deposited on agar medium (measured under a 10x objective, NA 0.45) and sealed microfluidic devices. Normally, calibrations require 4–5 known O_2 concentrations to cover the working range. Rough calibrations and recalibrations can be conducted with two O_2 standard levels—ambient air (20.86 %) and 0 % O_2. A number of calibrations for different models used in the biological studies and examples described in the following sections are summarised in Table 3.2.

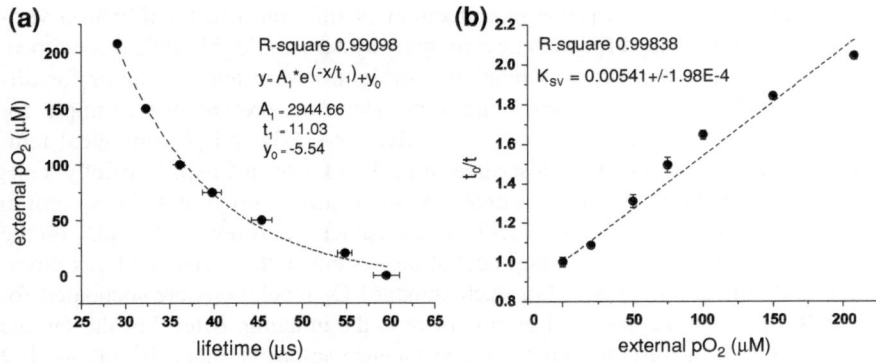

Fig. 3.2 Calibration of the NanO2 probe loaded in mammalian cells at 37 °C presented in lifetime (**a**) and Stern–Volmer (**b**) scale

Table 3.2 Calibrations of PtPFPP-based O_2 probes generated with different biological models and conditions

Biological System	LCI setup	Calibration Function
Mammalian cells (MEF) in DMEM, medium, 37 °C	35 mm MatTek dishes, adherent culture	$[icO2] = 2944_{*e}\left(\frac{\tau}{11.03}\right) - 5.54$
SK-N-AS spheres, DMEM, 37 °C	35 mm MatTek dish, culture of free-floating spheroids	$[icO2] = 2628_{*e}\left(\frac{\tau}{13.9}\right) - 29$
Biofilms of *P. aeruginosa* grown in Luria–Bertani (LB) medium, 23 °C	Microfluidic biochips (plastic)	$[icO2] = 33294_{*e}\left(\frac{\tau}{5.84}\right)$
C. elegans grown on agar plates with *E.coli* colonies, 23 °C	Agar plates with probe deposited on surface	$[icO2] = 8534_{*e}\left(\frac{\tau}{8.34}\right)$

3.9 Factors Affecting the Quality of Wide-Field FLIM Images

Digital images of live biological samples acquired on a wide-field FLIM microscope have to be corrected for several contributing factors originated from the sample parameters as well as from characteristics of the optical system and photon detector [27].

Non-specific *background signal* arises from: (i) out of focus events; (ii) auto-fluorescence of the specimen, imaging dish or biochip material; (iii) components of the medium; iv) emission of other luminescent stains bleeding to the detector. Out of focus background is common for wide-field microscopy as the whole specimen is excited and the light from the focal plane and out of focus light is collected through the objective. Deconvolution can be used to post-process wide-field images with strong out of focus background [28, 29]. Confocal and multi-photon systems overcome this problem, but they have their own limitations such as longer signal acquisition times [30].

Another factor is *random noise*, which determines the quality of measured signals, in the resulting images it can be expressed as signal-to-noise ratio, S:N. The Poisson noise is the variance of collected photons per pixel in repeated measurements, without the contribution of photobleaching or movement artefacts. Thermal noise due to thermal generation of electrons is eliminated in cooled CCD cameras for which the dominant source of noise is read noise. The latter means that measured voltage has a variance for the number of photons detected [31]. Photomultiplier tubes (PMT) use a signal amplifier which adds additional noise due to amplification processes. Unlike confocal, wide-field images cannot be corrected for the total noise, since it is not a constant. However, signal acquisition can be optimised to generate sufficient S:N and reduce the impact of noise on image quality. In intensity-based imaging of intracellular parameters, it is particularly important to quantify background signal outside the cell and correct for it. In FLIM the contribution of background signals is not so critical as lifetime is determined from luminescence decay. However, for best quality images, it is recommended to assess background signals, optimise S:N and correct for background.

The key optimisation parameter is *exposure time* which defines the time during which the signal is integrated by the detector. Longer exposure time increases the signal and improves S:N, but it is limited by detector saturation when photons reaching the detector cannot increase the signal further. In this case, linearity is lost and saturated images cannot be used for quantitative intensity or FLIM analysis. Detector efficiency is a measure of the percentage of photons reaching the detector that are counted and is usually supplied by the manufacture. CCD detectors have quantum efficiency up to 60–90 %, compared to 10–20 % for PMT detectors, however, PMT detectors usually have wider dynamic range and better S:N.

Optimization of microscope *optics* is extremely important. Performance of the two optical filters and dichroic mirrors determine the efficiency of excitation and collection of the luminophore emission which should be maximised, while minimising bleeding of parasitic excitation light and auto fluorescence into the emission channel. Photodamage of the probe and biological sample must also be considered. Luminophore brightness is the product of its emission quantum yield and molar extinction coefficient [16], and one should account for spectral characteristics of optical background generated by the sample and microscope and optical bands provided by the excitation and emission filters. Biological applications usually require short exposure times to minimise phototoxicity and photodamage. To generate sufficient signals at low exposure times, binning, a function to form a pixel of larger size from several pixel, increases the amount of collected photons but at the cost of lower image resolution.

Spherical aberration is caused by the mismatch between the refractive index of the lens and the specimen, e.g. using an immersion oil objective with samples in aqueous medium [32]. Spherical aberration increases with the distance from the coverslip. The use of correction collars, refractive index matching objectives (immersion oil or water) can largely eliminate the loss of intensity signal. Also probe signals have a second power dependence on the objective numeric aperture

(NA^2). NA of immersion water and oil objectives and air objective have theoretical maximum of 1.3, 1.5 and 1.0, respectively.

Non-uniform illumination is common for LED excitation systems resulting in uneven luminescent readings over the field of view. This can be corrected by normalising intensity images of samples-of-interest for an image of even fluorescent sample or blank. Uneven illumination has less effect on PLIM measurements as lifetime is independent of the intensity.

Overall, to achieve best O_2 imaging results it is necessary to assess the available microscopy setup with the chosen O_2 probe, determine and optimise the key factors and parameters that influence performance of phosphorescent images they produce.

3.10 Examples of O_2 Imaging Experiments

3.10.1 Cultures of Adherent Cells

Cells cultured in flasks, microplates or imaging dishes continuously consume O_2 dissolved in the medium. Although individual cells have rather low OCR, respiration of cell populations is significant and can cause self-deoxygenation. As described in Chap. 2, the level of cell oxygenation depends on many parameters including the type of cells, their metabolic status, density and uniformity of surface coverage, thickness of the medium layer above the cells, medium composition (viscosity and spectrum of metabolic substrates utilised by the cells), environmental parameters (partial oxygen pressure, temperature), sample geometry and conditions of mass exchange [33]. While the monitoring of in situ oxygenation of cells within the sample by optical O_2 imaging techniques is relatively straightforward, a comprehensive assessment of the contribution of all these parameters related to cell oxygenation is difficult.

Unlike the systems requiring high cell numbers, such as the analysis of cell populations on a TR-F reader (see Chap. 2), high-resolution O_2 imaging enables tracing local O_2 levels and gradients in individual cells, small clusters of cells and larger, even macroscopic specimens. Non-specific binding of the probe outside the cells has no influence on specific signals, thus allowing unambiguous data interpretation. Furthermore, detailed 3D O_2 maps can be reconstructed using laser-scanning confocal and MPM techniques [34, 35, 36].

3.10.1.1 Wide-Field FLIM

Figure 3.3 shows representative transmission, phosphorescence intensity images and lifetime images generated on the wide-field FLIM system described earlier with MEF cells (3×10^4/well grown over 24 h) at rest and after treatment with

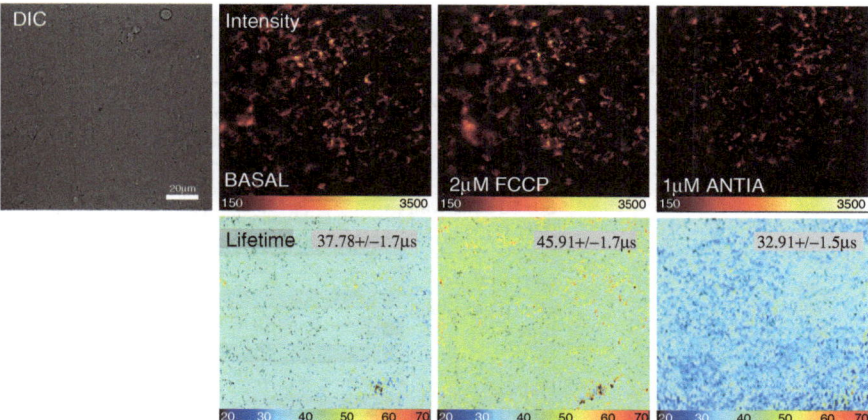

Fig. 3.3 Transmission, intensity images and lifetime images of MEF cells in basal respiration state and upon treatment with 2 μM FCCP (mitochondrial uncoupler) and 1 μM Ant A (complex III inhibitor), produced with NanO2 probe at ambient pO$_2$, 37 °C

FCCP and Ant A (activator and inhibitor of respiration, respectively). The lifetime images were generated over the whole field of view and converted into O$_2$ concentrations by applying the O$_2$ calibration. One can see that under ambient conditions (atmospheric pO$_2$ = 20.86 %) basal $\tau = 37.78 \pm 1.7$ μs corresponds to ~ 105 μM O$_2$ within the cell monolayer. This is much lower than the O$_2$ level in air-saturated medium (209 μM). Treatment with FCCP increased cellular respiration and reduced icO$_2$ over the entire field while inhibition of cellular respiration by Ant A increased cell oxygenation bringing it to ambient O$_2$ levels. The respiratory responses observed by the imaging technique correlated well with the results generated on TR-F plate reader for cell populations (see Chap. 2). The presented lifetime images underscore the robustness of FLIM modality. Even in out of focus areas, lifetime signals remained stable due to the fact that the phosphorescent signal was collected through the whole sample and that the lifetime and O$_2$ concentration were determined from the rate of the intensity decay rather than individual intensity values.

Monitoring a denser layer of MEF cells (6×10^4 cells/well) revealed a lower basal icO$_2$ concentration of ~ 60 μM. The addition of 500 nM valinomycin (K$^+$ ionophore and activator or respiration [37]) caused complete deoxygenation of the cells, and subsequent addition of 10 μM Ant A increased the icO$_2$ above the basal levels. HCT116, HepG2 and PC12 cells were measured in a similar way using the NanO2 probe. Figure 3.4 shows that these cells respond in a similar manner to drug treatment, but also show cell-specific local O$_2$ levels for each condition. Notably, in some of the cells the NanO2 probe was not distributed evenly producing 'patchy' patterns. This did not make imaging of cellular O$_2$ problematic, however, for the areas with low intensity signals fitting of the phosphorescence decay did not work well and non-fittable pixels had to be excluded from the resulting lifetime and icO$_2$ images.

Fig. 3.4 Average
oxygenation of different cells
(MEF, HCT116 and HepG2)
and cell densities under
resting and stimulated
conditions

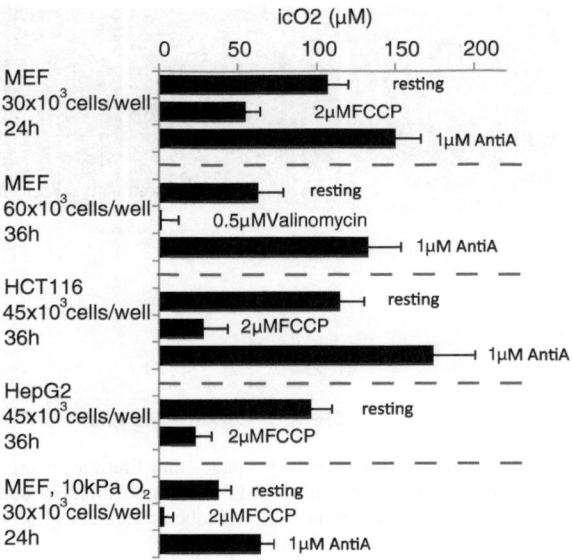

SK-N-AS neuroblastoma cells were analysed at a low density of $5*10^3$ cells/ well and different levels of atmospheric O_2. Figure 3.5 shows DIC, lifetime and O_2 maps for individual SK-N-AS cells at normoxia ($pO_2 = 18$ %, which corresponds to 180 μM in the medium) and hypoxia (2 % or 20 μM). The overlay of intensity and transmission image confirms that probe localises within the cells and does not bind non-specifically to the surface of imaging dishes. Several ROIs with individual and clusters of cells were defined and processed to determine average lifetimes and O_2 concentrations. In these conditions, measured icO_2 levels within the cells were close to the pre-set O_2 levels in the atmosphere and bulk of the medium. This means that the cell layer does not deoxygenate itself. Inside the cells icO_2 gradients could not be seen. This is due to the fast diffusion of O_2 across such small objects and/or significant measurement error (relative standard deviation was close to 10 %).

Oxygenation of an 80 % confluent layer of resting MEF cells was also analysed at different levels of atmospheric pO_2. Figure 3.6 shows a marked increase in *relative* deoxygenation (%) of the cell layer compared to bulk medium. Almost complete anoxia (>95 %) was reached already at about 5 % atmospheric O_2. It is worth noting that when studying adaptive responses to hypoxia, experiments with cells are normally conducted at 1–2 % atmospheric O_2 [33, 38], however, in many instances this would mean that the cells in the samples are actually exposed to complete anoxia.

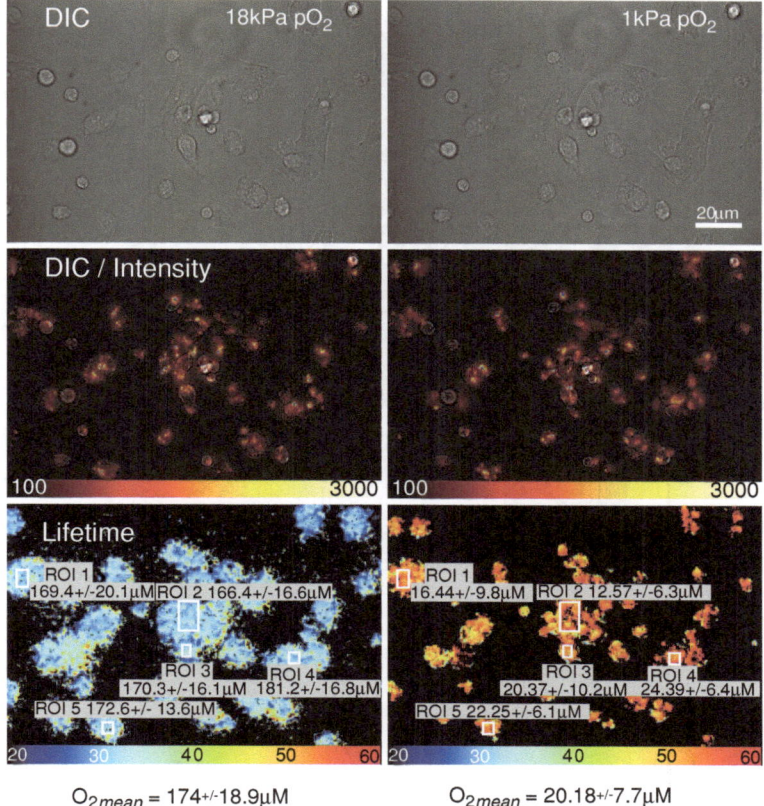

Fig. 3.5 Transmission, intensity and lifetime images of SK-N-AS cells, with calculated icO₂ concentrations in several ROI's and mean icO₂, at 18 % (*left panel*) and 2 % (*right panel*) of atmospheric O₂

Fig. 3.6 Relative deoxygenation of an 80 % confluent layer of resting MEF cells at different levels of atmospheric O₂

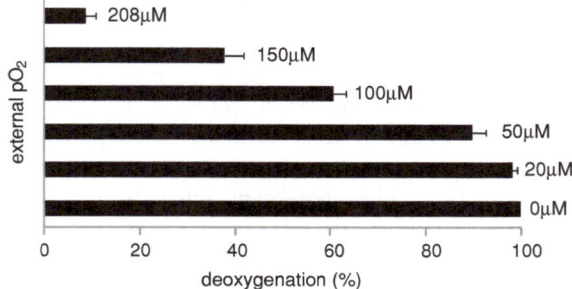

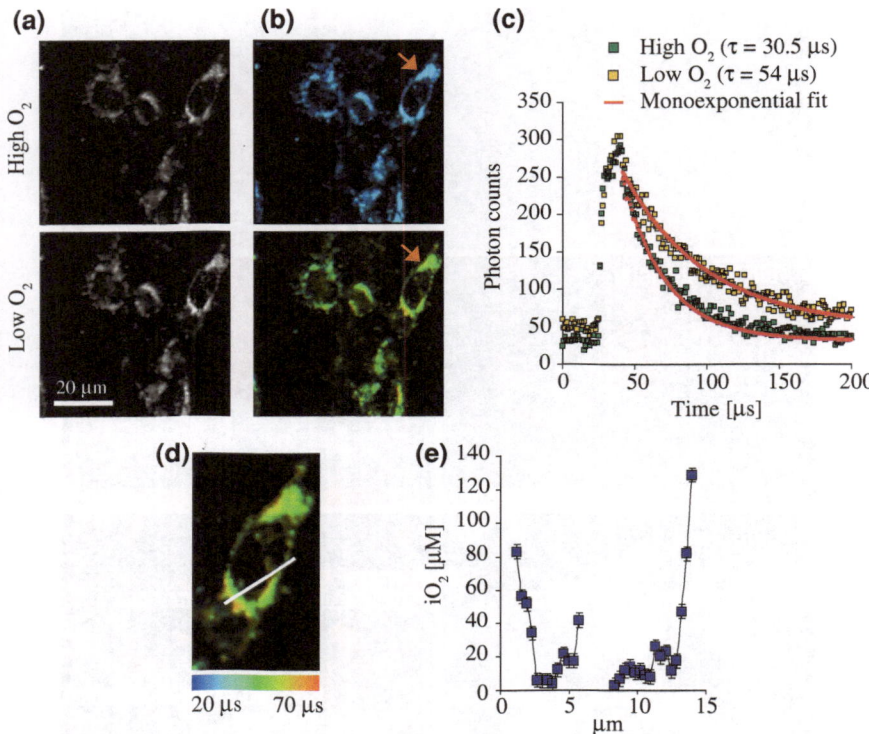

Fig. 3.7 Single plane confocal images of MEF cells loaded with NanO₂ probe (10 μg/ml), recorded in TCSPC mode. **a** Intensity profiles of the cells at high (20.86 %, *upper panel*) and low O₂ (*lower panel*, sulphite addition). **b** Corresponding lifetime images. **c** Representative phosphorescence decays and single exponential fits calculated for the ROI indicated by *arrow* in b, at high and low O₂. **d, e** Magnified lifetime image of the cell at low O₂ in psedo-colour scale and O₂ concentration profile across the cell (along the *white line*)

3.10.1.2 Confocal TCSPC-FLIM

To perform more detailed and precise analysis of spatial O₂ distribution within mammalian cell cultures a confocal laser-scanning FLIM system DCS-120 (Becker & Hickl, Berlin, Germany) was applied. MEFs were seeded at $1.5*10^4$ cells/cm², grown for 24 h in adherent state and stained with NanO2 probe (10 μg/ml) for 28 h. Phosphorescent intensity images (Fig. 3.7) were recorded as described above (Sect. 3.3), using a 405 nm laser and a 600 nm longpass filter. High brightness of the NanO2 probe provided rapid acquisition of intensity signals (A) and generation of 2D lifetime images (B, C). The lifetime data were converted to O₂ concentrations using the calibration equation: O₂ (μM) $= 65227.113*e^{(-\tau/5.73035)}$. One can see that at ambient O₂ levels, icO₂ distribution was even, however, after partial deoxygenation of medium with sulphite confocal FLIM revealed spatial differences in O_2 levels within the cells (D). Presumably these areas with reduced icO_2 concentrations are in close proximity to the mitochondria.

It is worth noting that the measurements presented in Fig. 3.7 were conducted under non-optimal temperature (~ 25 °C) and O_2 control (incubation chamber was not available on site). Therefore, this system requires further analysis and verification of the results, e.g. by co-staining with markers of cellular organelles and conducting more O_2 measurements 3D analysis under different conditions.

Overall, the above results illustrate that icO_2 probes such as NanO2 are useful for O_2 imaging and work reliably in individual cells under hypoxia, where precise control of cell oxygenation during both short-term and prolonged experiments is necessary. Also when O_2 approaches zero levels, phosphorescent probes show best accuracy and resolution, whereas electrochemical systems (Clark electrode) do not work reliably in this range.

3.10.2 Multicellular Spheroids and Neurospheres

The presence of 'cancer stem cells' has been reported in various solid tumours including brain, breast and prostate neoplasms [39]. Neuroblastoma, a neural crest derived from extra cranial solid tumour occurring in infants and children, has been shown to contain pluripotent tumour initiating cells [40]. Hypoxia is considered to promote reprogramming and de-differentiation of tumour cells leading to the emergence of a stem cell phenotype that is highly aggressive, able to evade chemotherapy and ultimately lead to poor clinical prognosis [41]. Also cancer stem cells can undergo self-renewal processes and form microscopic spheroids on non-adherent substrates in the presence of growth factors [42].

Neurosphere formation is also used to identify the presence of self-renewing cells, and to isolate stem cells of cerebral nerve system (CNS) [43]. Such 3D cell models represent useful research tools since they are physiologically more relevant than the commonly used 2D cell cultures [44].

3.10.2.1 Wide-Field Imaging of SK-N-AS Neurospheres

Since O_2 microenvironment within the neurosphere can contribute to stem cell and tumour development and determine tumour aggressiveness and 'stemness' of neuroblastoma, O_2 levels in neurospheres were investigated, initially using the spheres formed from SK-N-AS cell line, NanO2 probe and wide-field FLIM. The significant size and thickness (100–500 μm in diameter) of the neurospheres made focusing rather complicated. The focus was adjusted to obtain highest intensity signals, and an assumption was made that this corresponded to the centre of the sphere. Compared to the monolayer cultures of SK-N-AS cells, the spheres loaded with NanO2 probe showed approximately 5-fold greater intensity signals at ambient O_2 (18 %). This can be attributed to their 3D structure, high cell density a considerable depletion of O_2 inside the sphere. Oxygenation of the neurospheres and internal O_2 gradients were studied under normoxic and hypoxic conditions and measurement parameters were adjusted to maintain the same intensity signal at lower O_2 levels. Sets of intensity images of the neurospheres at different delay times were produced, converted into lifetime images and then into O_2 maps. This was done at three different levels of atmospheric O_2.

The images in Fig. 3.8 show that at 18 % O_2 the outer layer of the spheres ($\sim$ 10 μm) O_2 levels is equal to the external O_2, while deeper inside the sphere O_2 was markedly reduced. At 8 % O_2, a pronounced O_2 gradient was seen inside the spheres, while at 3 % it vanished since the whole sphere became completely deoxygenated. Relative deoxygenation of the central part of the sphere was 20 %, 90 % and 100 % at 18 %, 8 % and 3 % O_2, respectively. These maps reflect O_2 gradients starting from ambient O_2 concentration at the outer layer of the sphere, reaching a minimum of 120 μM, 15 μM and practically zero at the centre, respectively. This data prove that significant heterogeneity and O_2 gradients do exist in neuroblastoma spheres and that O_2 concentration in the surrounding medium has a profound effect on them. At the same time, wide-field images and O_2 maps in Fig. 3.8 represent mean values averaged across the Z-axis. Precise reconstruction of 3-D distribution of O_2 inside the spheres was not possible with this technique.

3.10.2.2 Confocal TCSPC-FLIM of Primary Cortical Neurospheres

To obtain a detailed 3D view of O_2 distribution within the neurophere structure, similar experiments were conducted on the DCS-120 confocal TCSPC-FLIM microscope described above. In this case, neurosphere culture was prepared from the E18 rat embryo brain cortexes. All the procedures with animals were performed under licence issued by the Department of Health and Children (Ireland) and in accordance with the European Community Council Directive (86/609/EEC). Neurons isolated from 10–14-day rat brain embryos were grown for 7 days in DMEM/F12 Ham medium supplemented with 2 % B27, 20 ng/ml of EGF and FGF, penicillin and streptomycin [45]. The neurospheres were loaded during their formation, by adding 20 μg/ml NanO2 to the medium on days 2, 4 and 6. Prior to the imaging, neurospheres (50–200 μm in diameter) were immobilized on glass bottom dishes (MatTek, MA) coated with poly-D-lysine.

Phosphorescence intensity images of a single plane cross-section of the neurosphere and corresponding lifetime image produced by the analysis of the phosphorescence decays for each pixel in SPCImage software (Becker&Hickl) are shown in Fig. 3.9. Lifetime data were converted to O_2 concentrations as above (see Sect. 3.10.1.2). Line profile (X–Y) analysis demonstrated large difference in O_2 levels between the cells residing in the inner (deoxygenated) and outer (more oxygenated) regions of neurospheres (D). Similar changes were observed along Z-axis when analysing serial cross-sections: O_2 levels progressively decreased accordingly to the distance from the top to the centre of the neurosphere (E).

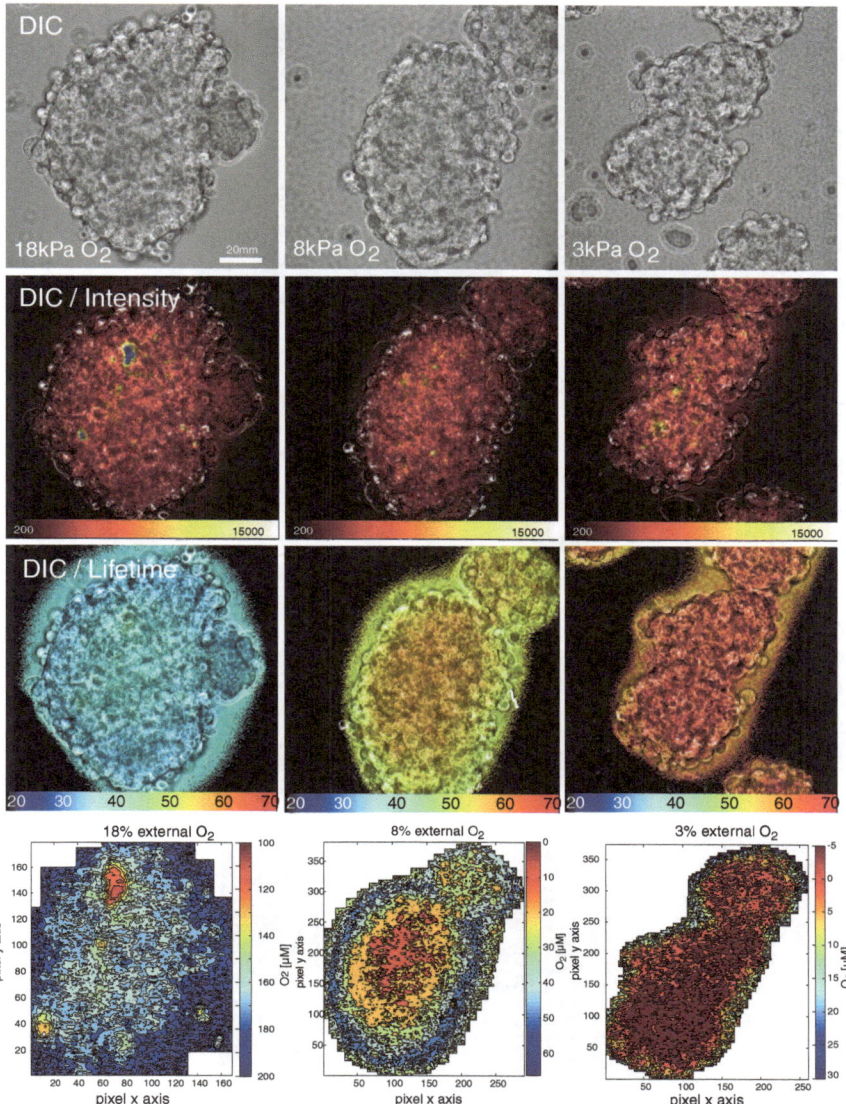

Fig. 3.8 Wide-field transmission, phosphorescence intensity and lifetime images and reconstructed O$_2$ maps for SK-N-AS neurospheres at 18 %, 8 % and 3 % of atmospheric O$_2$

Finally, when mitochondrial respiration was inhibited by addition of Ant A, O$_2$ concentrations in neurospheres gradually increased. Thus, treatment of the spheres for 15 min caused an increase in local O$_2$ level from 2 to 6 µM (F).

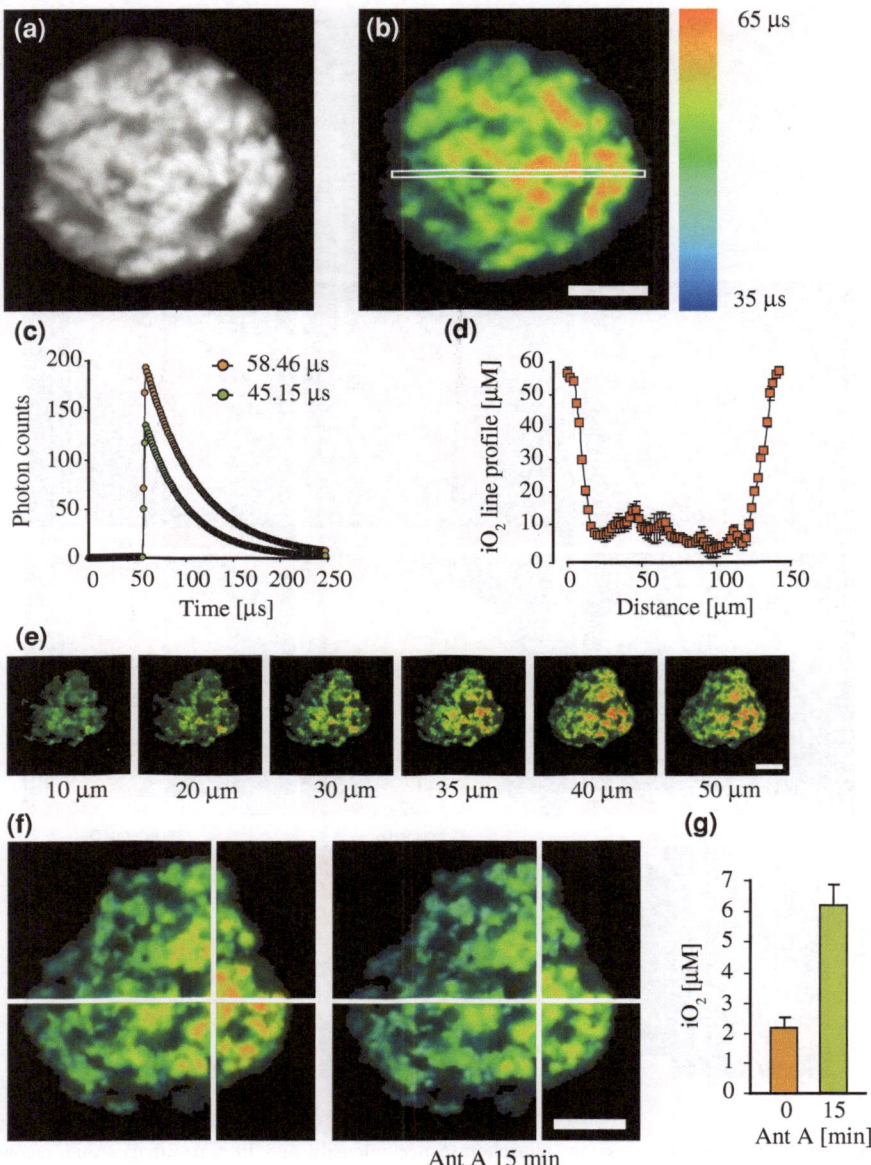

Fig. 3.9 Confocal TCSPC-FLIM-based O$_2$ imaging of cortical neurospheres. **a** Phosphorescence intensity image of the neurosphere cross-section. **b** Lifetime image derived from (a). Colour spectrum shows the range of measured lifetimes. **c** Probe phosphorescence decays for two pixels with $\tau = 45.15$ μs and $\tau = 58.46$ μs. **d** O$_2$ profile across a single confocal stack (indicated in (b) by *rectangular*) (averaged 5×1 pixel). **e** O$_2$ images of nurosphere cross-sections (in μm). **f** Ant A (10 μM) gradually increases O$_2$ levels in the neurosphere. **g** At the point of line intersections in (f), O$_2$ increased from 2 to 6 μM. Scale bar—50 μm

3.10.2.3   Two-Photon Ratiometric O_2 Imaging of Primary Cortical Neurospheres

Despite its advantages over the wide-field microscopy, confocal fluorescence microscopy has its own limitations, particularly when applied to the imaging of live tissues and organisms. Illumination-related tissue damage and limited penetration of the laser beam can be improved significantly by using multi-photon microscopy. The recently developed nanoparticle-based probe MM2 [26], which is compatible with two-photon laser excitation and allows ratiometric detection due to second antennae dye not present in the NanO2 probe was applied to the analysis of O_2 distribution in cortical neurospheres.

Experiments were performed on a multiphoton microscope Olympus FV1000 MPE equipped with Mai Tai® DeepSeeTM Ti:Sapphire laser (Newport Corporation, CA) in the photon counting mode. The presence of O_2-sensitive PtTFPP and O_2-non-sensitive antennae dye (poly(9,9-diheptylfluorene-alt-9,9-di-p-tolyl-9H-fluorene)) (PFO) in MM2 particles allows for ratiometric detection of the changes in cell/tissue oxygenation. Following the neurosphere preparation and loading (performed as described above), the probe was excited at 760 nm (optimal wavelength) and emission was collected in the two spectral windows: 420–460 nm for O_2-insensitive PFO and 605–680 nm for O_2-sensitive PtTFPP dyes. Multi-photon microscopy at 20 μs/pixel scanning speed revealed that the signals of the two dyes co-localize inside the neurospheres, and the ratio of the PtTFPP and PFO signals was higher in the inner regions of neurosphere, thus reflecting their reduced oxygenation. Deoxygenation of the medium with sulphite increased the ratio (Fig. 3.10).

These results agree with the results produced by the wide-field and confocal FLIM imaging of neurospheres using the NanO2 probe. Overall, they demonstrate broad applicability of the nanoparticle-based icO2 probes for O_2 imaging in complex 3D biological objects such as neurospheres. Their ease of use combined with high brightness and photostability provide good analytical performance in O_2 imaging experiments using a number of different microscopy techniques. They can be used in in vivo studies of physiological responses, effects if hypoxia, stress and metabolic perturbations.

3.10.3   Microfluidic Systems and Cell-Based Biochips

Cell based microfluidic and lab-on-a-chip systems are gaining broader use in biomedical research, including long-term cell cultures, stem cell proliferation and differentiation [46], studies with 3D biological and disease models, process and media optimisation and drug screening. Compared to conventional culture systems, microchamber arrays allow rapid, high throughput, multi-parametric biological analyses, large reduction in reagents and labour costs, improved data quality and automation [47]. They can facilitate a better understanding of complex interactions and, processes occurring within the cell, cellular responses to

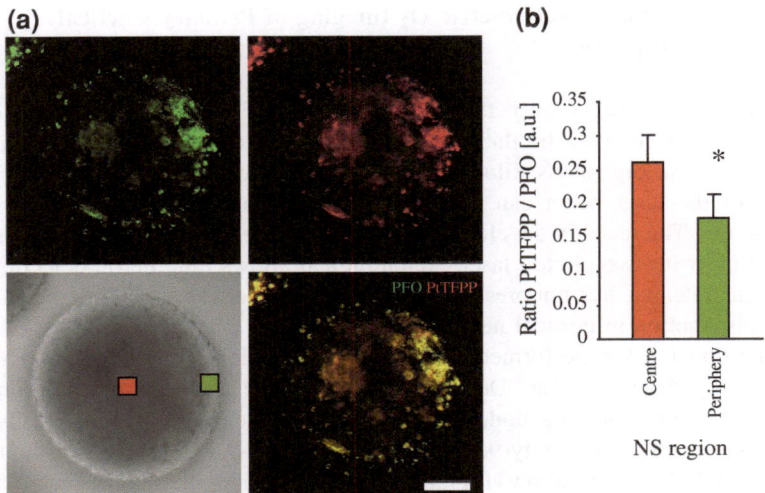

Fig. 3.10 Ratiometric intensity imaging of cortical neurospheres using multi-photon microscopy and MM2 probe. **a** Localization of PFO and PtTFPP signals inside the neurosphere. **b** PtTFPP/PFO intensity ratio calculated for the inner and outer regions, indicated on the bright-field image (a). Bar represents 100 μm. Asterisk shows statistical difference

changing conditions, and can be coupled to different detection platforms including live-cell imaging.

To avoid evaporation of small volumes of liquid and prevent formation of bubbles and air plugs, biochips usually operate as a sealed (or partially sealed) system in a flow mode. In such biochips containing respiring cells control of oxygenation becomes critical, since inadequate or interrupted supply of O$_2$ induced by stopped flow or overgrown cells can quickly suffocate the cells inside the chip, thus killing them or altering their normal function. On the other hand, microfluidic systems can be used to create and maintain O$_2$ gradients [38]. Optical O$_2$ sensing with intra- and extracellular phosphorescent probes provides a simple technique to control oxygenation conditions inside the biochip, non-invasively, on a microscale and with imaging capability.

Using the microfluidic system developed at the Department of Micro- and Nanotechnology, Technical University of Denmark, DTU Nanotech [48] and the above wide-field CCD camera-based FLIM system, O$_2$ levels inside microchambers containing a cultured biofilm of *Pseudomonas aeruginosa* were monitored. These are a prokaryotic cells which grow under micro-aerobic conditions and microchambers were inoculated with laboratory strain PAO1 of *P.aeruginosa*, maintained for 12 h under constant flow conditions (8.5 μL/min) in the medium containing the NanO2 probe (10 μg/ml), then washed with fresh medium and analysed under the microscope. It was found that microbial biofilms, after they were exposed to the O$_2$ probe in the microchambers, became stained producing strong phosphorescent signals (Fig. 3.11), and biochips with such microbial films

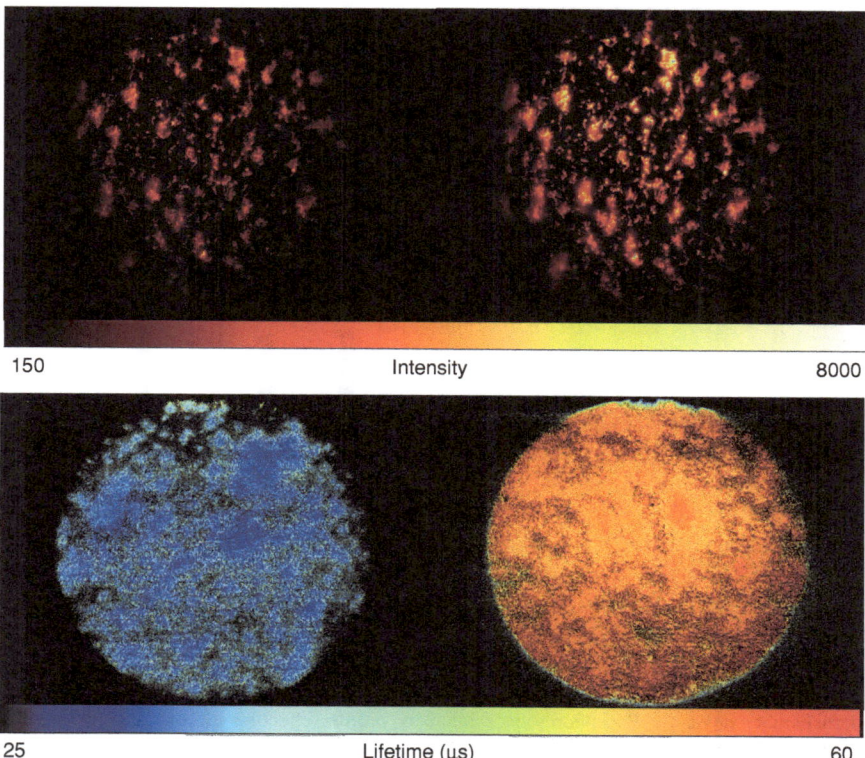

150   Intensity   8000

25   Lifetime (µs)   60

Fig. 3.11 Phosphorescence intensity and lifetime images of P. aeruginosa biofilm in a microfluidic chip under aerated (fast flow rate, 20.8 % O$_2$) and hypoxic (stopped flow, 0 % O$_2$) conditions

can be used to conduct biological experiments for several days. When measured under the aerobic (fast flow rate), micro-aerobic (slow flow rate) and anaerobic (stopped flow) conditions, average lifetime values were determined to be 30.5 µs, 35.8 µs and 51.1 µs, which corresponds to 94.2 %, 62.7 % and 7.9 % of air-saturation within the biofilm and microchamber, respectively.

In another experiment MEF cells were grown in standard 75 cm^2 flasks, then trypsinised and seeded in the chambers of the microfluidic biochip at a concentration of 1x10^6 cells/ml in the presence of 1 mg/ml microparticle-based probe comprising 3 µm PS-DVB microparticles impregnated with 1 % PtBP. After 1 h incubation under stopped flow conditions, attachment of the cells as well as the O$_2$ microparticle sensors to the surface of the chip was observed. Corresponding wide-field and fluorescent images of the internal area of the microchamber are shown in Fig. 3.12.

After that, focusing the microscope on the cell surface, O$_2$ levels inside the microchambers with cells were monitored by wide-field O$_2$ imaging technique under different flow conditions. Phosphorescence intensity profile in Fig. 3.12b

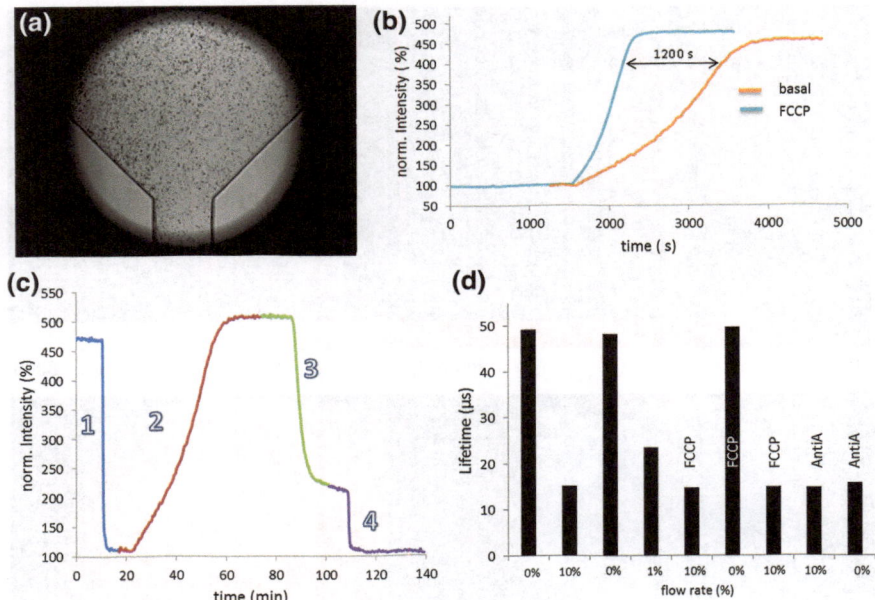

Fig. 3.12 a Transmission image of the biochip microchamber loaded with MEF cells (*grey dots*) and the PtBP-based microsensors (*black dots*). **b** Deoxygenation profiles of resting (*orange curve*) and FCCP uncoupled (*blue curve*) MEF cells produced under stopped flow condition in the microchamber device. **c** Phosphorescence intensity profile reflecting the oxygenation of MEF cells under different flow conditions: stopped flow—high flow rate–low flow rate–high flow rate. **d** Average lifetime values reflecting microchamber oxygenation levels under different flow conditions and pharmacological treatments (FCCP and Ant A). Measured on the wide-filed O$_2$ imaging system using PtBP microparticles (extracellular probe)

illustrates that under stopped flow condition at the start the microchamber was completely deoxygenated due to cell respiration in a closed volume, and when the flow was resumed at high rate (10 %, corresponds to the 8.5 µL/min) rapid oxygenation of the microchamber occurred and dissolved O$_2$ concentration was brought to air-saturating levels (Zone 1). When the flow was stopped again, the process of O$_2$ depletion resumed (Zone 2). Its kinetics was followed by the O$_2$ sensing, from which basal OCR of MEF cells can be determined (processed slope [49]). After the cells reached anoxia, the flow was resumed again at a lower rate (0.85 µL/min). Under this condition the microchamber remained partly deoxygenated (Zone 3), and increasing the flow rate brought the O$_2$ to ambient levels (Zone 4). Such deoxygenation and reoxygenation cycles can be repeated many times, and lifetime values for plateau regions on Fig. 3.12c show good reproducibility of the microfluidic system. Treatment of cells with drugs (FCCP and Ant A), acute or sustained hypoxia can also be conducted and OCRs for these conditions measured and compared with resting normoxic cells.

3.10.4 O_2 Gradients in Cultures of *C. elegans* Worms

Caenorhabditis elegans is a free-living, nonparasitic, multicellular metazoan with a short life cycle of approximately 3d that has been used as a model organism in developmental biology, behaviour, anatomy and genetics laboratories [50]. *C. elegans* is a common model for neuroscience research as their neuron system is well described and established [51], but still some of their neuronal functions, physiological behaviour and responses to O_2 availability are poorly understood. OCR of small numbers of *C. elegans* can be measured on a plate reader and in capillary cuvettes with individual animals on the LightCyler® system [52]. On the other hand, O_2 gradients surrounding individual and groups of *C. elegans* (normally cultured in air atmosphere on agar plates fed with *E.coli* colonies) can be studied by O_2 sensing and imaging techniques, particularly FLIM.

C.elegans were cultured under standard conditions and transferred via a sterilised transport needle into the centre of agar plate which was pre-soaked with concentrated stock of NanO2 probe (5 mg/ml). The plate with worms was placed upside down in the incubation chamber of the microscope set at room temperature (23 °C). Since imaging of live *C.elegans* was affected by their rapid movement, imaging conditions were optimised to eliminate measurement artefacts, such as differences in DIC and lifetime images.

Using a FLIM microscope comprising a simple manual system on which the DIC and FLIM imaging modes are used sequentially, the following strategy was used. Probe concentration was increased (see above) to maximise the intensity signals, while integration time and the number of frames in FLIM data acquisition were reduced to a minimum. This allowed lifetime images to be taken in less than 500 ms ($\sim$50 ms per frame) over the whole field. Examples of such images are shown in Fig. 3.13, in which one can see good co-localisation of the O_2 gradient with the clamp of *C. elegans*. Although some movement artefacts are still seen, and the initial DIC image does not exactly match the subsequent intensity image(s). Further improvement can be achieved by using *C.elegans* pharmacologically paralysed for a short period of time (several minutes). However, this should be done with care since animal respiration and O_2 gradients can also change significantly as a result of such treatment.

This example illustrates the potential of optical O_2 sensing and FLIM techniques with respect to terrestrial invertebrate models and measurement of O_2 gradients at the interface of solid phase (agar or soil) and gaseous atmosphere. Using such models, oxygenation of organisms, formations of clusters and roles of O_2 availability can be studied and related to more complex models such as higher mammals.

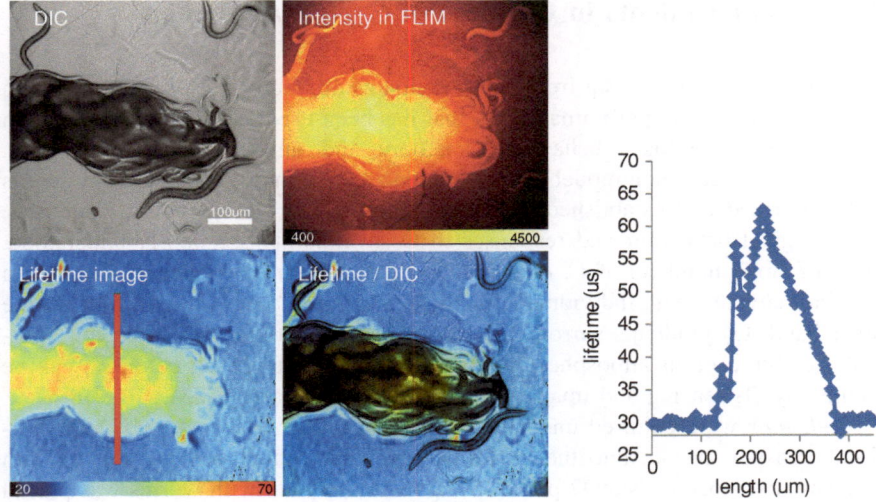

Fig. 3.13 Imaging of live *C.elegans* worms on agar plates and O₂ gradients surrounding them, under 21 % atmospheric O₂, 23 °C

3.11 Conclusions

Overall, this chapter gives a comprehensive description of the fundamentals of O₂ imaging techniques based on quenched-phosphorescence detection of Pt-porphyrin probes, particularly intracellular probes. Basic theoretical and practical considerations underpinning this powerful experimental methodology, different instrumentation and detection options are described and cross-compared. The most critical factors that determine the performance of imaging experiments with respiring specimens, rationale for instrument selection and optimisation of measurement settings are highlighted. Representative examples, in which different biological models, including adherent cell cultures, multicellular spheroid systems (neurospheres), *C. elegans* worms and cell-based microfluidic biochips are provided. In these case studies the different uses of O₂ imaging and specific bioanalytical tasks and imaging formats are described in greater detail and corresponding experimental results are explained and discussed. All this illustrates the broad scope of practical uses of these techniques, their analytical potential, particularly in the studies of cellular function, metabolism, cell and tissue physiology, hypoxia research, roles of O₂ in biological systems.

Acknowledgments This work was supported by the Science Foundation Ireland, grant 07/IN.1/ B1804, EU projects FP7-MC-IAPP-2009-230641, FP7-HEALTH-2012-INNOVATION-304842-2. Authors wish to thank the researchers who helped with experiments and results presented in this chapter, particularly Dr. Ruslan I. Dmitriev (Biochemistry Department, University College Cork); Dr. Violine See and Anne Herrmann (Institute of Integrative Biology University of Liverpool, UK)— wide-field imaging of SK-N-AS neurospheres; Dr. Yvonne Nolan and Ian O'Brien (Anatomy department, University College Cork)—isolation of primary neurons from rat brain; Dr. Wolfgang Becker and Axel Bergmann (Becker & Hickl GmbH, Germany)—confocal TCSCP-FLIM measurements; Mr. Zoltan Soltesz and Dr. Mario de Bono (Medical Research Council Laboratory of Molecular Biology, University of Cambridge, UK)—wide-field imaging of *C. elegans;* Dr. Maciej Skolimowski, Prof. Jenny Emneus (Department of Micro- and Nanotechnology, Technical University of Denmark, Copenhagen), Mr. Norbert Galler (Graz University of Technology, Austria)—measurements in microfluidic biochips; Dr. Sergei Borisov (Institute of Analytical Chemistry and Food Chemistry, Graz University of Technology, Austria)—synthesis of the microparticle probe.

References

1. Papkovsky DB (2010) Live cell imaging: methods and protocols, Meth Mol Biol, vol 591. Humana Press, 367p
2. Nägerl UV, Willig KI, Hein B, Hell SW, Bonhoeffer T (2008) Live-cell imaging of dendritic spines by STED microscopy. Proc Nat Acad Sci 105:18982–18987
3. Wenus J, Dussmann H, Paul P, Kalamatianos D, Rehm M, Wellstead P, Prehn J, Huber H (2009) ALISSA: an automated live-cell imaging system for signal transduction analyses. Biotechniques 47:1033–1040
4. Rehm M, Huber HJ, Hellwig CT, Anguissola S, Dussmann H, Prehn JH (2009) Dynamics of outer mitochondrial membrane permeabilization during apoptosis. Cell Death Differ 16:613–623
5. Carlin LM, Makrogianneli K, Keppler M, Fruhwirth GO, Ng T (2010) Visualisation of signalling in immune cells. Methods Mol Biol 616:97–113
6. Wouters FS, Verveer PJ, Bastiaens PIH (2001) Imaging biochemistry inside cells. Trends Cell Biol 11:203–211
7. Lippincott-Schwartz J (2011) Emerging in vivo analyses of cell function using fluorescence imaging. Annu Rev Biochem 80:327–332
8. Wilt BA, Burns LD, Wei Ho ET, Ghosh KK, Mukamel EA, Schnitzer MJ (2009) Advances in light microscopy for neuroscience. Ann Rev Neurosci 32:435–506
9. Becker W, Bergmann A, Biskup C (2007) Multispectral fluorescence lifetime imaging by TCSPC. Microsc Res Tech 70:403–409
10. Devor A, Sakadzic S, Srinivasan VJ, Yaseen MA, Nizar K, Saisan PA, Tian P, Dale AM, Vinogradov SA, Franceschini MA, Boas DA (2012) Frontiers in optical imaging of cerebral blood flow and metabolism. J Cereb Blood Flow Metab
11. Bastiaens PIH, Squire A (1999) Fluorescence lifetime imaging microscopy: spatial resolution of biochemical processes in the cell. Trends Cell Biol 9:48–52
12. Dmitriev RI, Papkovsky DB (2012) Optical probes and techniques for O_2 measurement in live cells and tissue, Cell Mol Life Sci 69(12):2025–2039
13. Dunphy I, Vinogradov SA, Wilson DF (2002) Oxyphor R2 and G2: phosphors for measuring oxygen by oxygen-dependent quenching of phosphorescence. Anal Biochem 310:191–198
14. Vanderkooi JM, Maniara G, Green TJ, Wilson DF (1987) An optical method for measurement of dioxygen concentration based upon quenching of phosphorescence. J Biol Chem 262:5476–5482
15. Ast C, Schmälzlin E, Löhmannsröben H-G, van Dongen JT (2012) Optical Oxygen Micro- and Nanosensors for Plant Applications. Sensors 12:7015–7032
16. Lakowicz JR, Masters BR (2008) Principles of fluorescence spectroscopy. J Biomed Optics 13:029901–029902
17. Liebsch G, Klimant I, Frank B, Holst G, Wolfbeis OS (2000) Luminescence lifetime imaging of oxygen, pH, and carbon dioxide distribution using optical sensors. Appl Spectrosc 54:548–559

18. Fercher A, O'Riordan TC, Zhdanov AV, Dmitriev RI, Papkovsky DB (2010) Imaging of cellular oxygen and analysis of metabolic responses of mammalian cells. Methods Mol Biol 591:257–273
19. Becker W, Su B, Holub O, weisshart K. (2010) FLIM and FCS detection in laser-scanning microscopes: Increased efficiency by GaAsP hybrid detectors. Microscop Res Tech 74:804–811
20. Periasamy A, Diaspro A (2003) Multiphoton microscopy. J Biomed Opt 8:327–328
21. Mills JD, Stone JR, Rubin DG, Melon DE, Okonkwo DO, Periasamy A, Helm GA (2003) Illuminating protein interactions in tissue using confocal and two-photon excitation fluorescent resonance energy transfer microscopy. J Biomed Opt 8:347–356
22. Lecoq J, Parpaleix A, Roussakis E, Ducros M, Houssen YG, Vinogradov SA, Charpak S (2011) Simultaneous two-photon imaging of oxygen and blood flow in deep cerebral vessels. Nat Med 17:893–898
23. Sakadzic S, Roussakis E, Yaseen MA, Mandeville ET, Srinivasan VJ, Arai K, Ruvinskaya S, Devor A, Lo EH, Vinogradov SA, Boas DA (2010) Two-photon high-resolution measurement of partial pressure of oxygen in cerebral vasculature and tissue. Nat Meth 7:755–759
24. Patterson MS, Madsen SJ, Wilson BC (1990) Experimental tests of the feasibility of singlet oxygen luminescence monitoring in vivo during photodynamic therapy. J Photochem Photobiol, B 5:69–84
25. Fercher A, Borisov SM, Zhdanov AV, Klimant I, Papkovsky DB (2011) Intracellular O2 sensing probe based on cell-penetrating phosphorescent nanoparticles. ACS Nano 5:5499–5508
26. Kondrashina AV, Dmitriev RI, Borisov SM, Klimant I, O'Brian I, Nolan YM, Zhdanov AV, Papkovsky DB (2012) A phosphorescent nanoparticles based probe for sensing and imaging of (intra)cellular oxygen in multiple detection modalities. Adv Funct Mater. doi:10.1002/adfm.201201387 (in press)
27. Waters JC (2009) Accuracy and precision in quantitative fluorescence microscopy. J Cell Biol 185:1135–1148
28. Murray JM (2011) Methods for imaging thick specimens: confocal microscopy, deconvolution, and structured illumination. Cold Spring Harbor Protocols 2011, pdb.top066936
29. Sarder P, Nehorai A (2006) Deconvolution methods for 3-D fluorescence microscopy images. Signal Process Mag, IEEE 23:32–45
30. Mertz J (2011) Optical sectioning microscopy with planar or structured illumination. Nat Meth 8:811–819
31. Pawley JB (1994) Sources of noise in three-dimensional microscopical data sets. In: Three-dimensional confocal microscopy: volume investigation of biological specimens (J.K. Stevens, L. R. M., J.E. Trogadis, Ed.), pp 47–94, Academic Press, San Diego
32. Huang B, Jones SA, Brandenburg B, Zhuang X (2008) Whole-cell 3D STORM reveals interactions between cellular structures with nanometer-scale resolution. Nat Meth 5:1047–1052
33. Zhdanov AV, Ogurtsov VI, Taylor CT, Papkovsky DB (2010) Monitoring of cell oxygenation and responses to metabolic stimulation by intracellular oxygen sensing technique. Integr Biol 2:443–451
34. Finikova OS, Lebedev AY, Aprelev A, Troxler T, Gao F, Garnacho C, Muro S, Hochstrasser RM, Vinogradov SA (2008) Oxygen microscopy by two-photon-excited phosphorescence. Chem Phys Chem 9:1673–1679
35. Sakadzic S, Roussakis E, Yaseen MA, Mandeville ET, Srinivasan VJ, Arai K, Ruvinskaya S, Wu W, Devor A, Lo EH, Vinogradov SA, Boas DA (2011) Cerebral blood oxygenation measurement based on oxygen-dependent quenching of phosphorescence. J Vis Exp e1694.
36. Wilson DF, Finikova OS, Lebedev AY, Apreleva S, Pastuszko A, Lee WM, Vinogradov SA (2011) Measuring oxygen in living tissue: intravascular, interstitial, and "tissue" oxygen measurements. Adv Exp Med Biol 701:53–59

37. Zhdanov AV, Dmitriev RI, Papkovsky DB (2010) Bafilomycin A1 activates respiration of neuronal cells via uncoupling associated with flickering depolarization of mitochondria. Cell Mol Life Sci 68:903–917

38. Lo JF, Wang Y, Blake A, Yu G, Harvat TA, Jeon H, Oberholzer J, Eddington DT (2012) Islet preconditioning via multimodal microfluidic modulation of intermittent hypoxia. Anal Chem 84:1987–1993

39. Das S, Srikanth M, Kessler JA (2008) Cancer stem cells and glioma. Nat Clin Pract Neuro 4:427–435

40. Mahller YY, Williams JP, Baird WH, Mitton B, Grossheim J, Saeki Y, Cancelas JA, Ratner N, Cripe TP (2009) Neuroblastoma cell lines contain pluripotent tumor initiating cells that are susceptible to a targeted oncolytic virus. PLoS One 4:e4235

41. Heddleston JM, Li Z, McLendon RE, Hjelmeland AB, Rich JN (2009) The hypoxic microenvironment maintains glioblastoma stem cells and promotes reprogramming towards a cancer stem cell phenotype. Cell Cycle 8:3274–3284

42. Reynolds BA, Weiss S (1992) Generation of neurons and astrocytes from isolated cells of the adult mammalian central nervous system. Science 255:1707–1710

43. Pevny L, Rao MS (2003) The stem-cell menagerie. Trends Neurosci 26:351–359

44. Jensen JB, Parmar M (2006) Strengths and limitations of the neurosphere culture system. Mol Neurobiol 34:153–161

45. Keohane A, Ryan S, Maloney E, Sullivan AM, Nolan YM (2010) Tumour necrosis factor-α impairs neuronal differentiation but not proliferation of hippocampal neural precursor cells: Role of Hes1. Mol Cell Neurosci 43:127–135

46. Meyvantsson I, Beebe DJ (2008) Cell culture models in microfluidic systems. In: Annual review of analytical chemistry, Annual Reviews, Palo Alto, pp 423–449

47. El-Ali J, Sorger PK, Jensen KF (2006) Cells on chips. Nature 442:403–411

48. Skafte-Pedersen P, Hemmingsen M, Sabourin D, Blaga FS, Bruus H, Dufva M (2012) A self-contained, programmable microfluidic cell culture system with real-time microscopy access. Biomed Microdevices 14:385–399

49. Ogurtsov VI, Hynes J, Will Y, Papkovsky DB (2008) Data analysis algorithm for high throughput enzymatic oxygen consumption assays based on quenched-fluorescence detection. Sens Actuators B: Chem 129:581–590

50. Wheelan SJ, Boguski MS, Duret L, Makałowski W (1999) Human and nematode orthologs: lessons from the analysis of 1800 human genes and the proteome of Caenorhabditis elegance. Gene 238:163–170

51. Dimitriadi M, Hart AC (2010) Neurodegenerative disorders: insights from the nematode caenorhabditis elegance. Neurobiol Dis 40:4–11

52. Zitova A, Hynes J, Kollar J, Borisov SM, Klimant I, Papkovsky DB (2010) Analysis of activity and inhibition of oxygen-dependent enzymes by optical respirometry on the Light Cycler system. Anal Biochem 397:144–151